PRAISE FOR *WHY?*

"Have you ever wondered why we wonder why? Mario Livio has, and he takes you on a fascinating quest to understand the origin and mechanisms of our curiosity. I thoroughly recommend it."

—Adam Riess, Nobel Prize winner in Physics, 2011

"An energetic look at the psychology and neuroscience of our inquisitiveness."

—Dan Jones, *Nature*

"In *Why?,* astrophysicist Mario Livio argues that humans are the only species to ask not just what, where or who, but also why. . . . His research roves broadly, from historical documents and technical studies to personal interviews. . . . Includes some fascinating tidbits along the way."

—Katherine Harmon Courage, *The Washington Post*

"A delightful romp through every aspect of human curiosity. It will surprise you, make you smarter, and put a spring in your step."

— Steven Strogatz, Jacob Gould Schurman Professor of Applied Mathematics, Cornell University, and author of *The Joy of X*

"A lively, expert, and definitely not dumbed down account of why we're curious."

—*Kirkus Reviews*

"[Livio] investigates the different shapes curiosity can take, how it expresses itself and the regions of the brain in which it appears to reside."

—*The New York Times*

"This cogent book presents the scientific research on curiosity in understandable ways without too much jargon."

—Joseph Peschel, *Science*

"An intellectual feast for any curious person."

—Jeffrey M. Schwartz, MD, research psychiatrist, UCLA, and coauthor of *The Mind and the Brain* and *You Are Not Your Brain*

"Whether in science or art, curiosity is essential to progress—but what is it, exactly? . . . Livio's book doesn't pretend to have all the answers, but it might well spur your own curiosity."

—David Lindley, author of *Uncertainty* and *Where Does the Weirdness Go*

"Well worth reading simply for the breadth of information about the subject, but also provides tips and cues readers may use to increase their own level of curiosity."

—Christopher M. Doran, *New York Journal of Books*

"A spellbinding journey through the latest findings on curiosity in psychology and neuroscience. Anybody who is curious about curiosity will want to read this book."

—Francesca Fiorani, creator of the digital archive Leonardo da Vinci and his Treatise on Painting and associate dean for the Arts and Humanities, University of Virginia

"A colorful, engaging, and yet thorough examination of curiosity in its many forms, *Why?* illustrates how having an inquisitive mind is crucial to creative output in the arts, sciences, and business. . . . A fascinating and fun read!"

—Gail Saltz, MD, clinical associate professor of psychiatry, The New York Presbyterian Hospital; TV commentator; host of "The Power of Different" podcast; and author of *The Power of Different*

"Brilliant."

—*Baltimore City Paper*

"[*Why?*] has a much wider appeal because Livio's vision is broad. He brings in literature, art and history and, crucially, the personal experience of a range of people who themselves are extremely curious. His style is engaging and playful, so even if you're not that much into science, you'll find plenty of ideas in this book to keep you thinking."

—*Plus Magazine*

ALSO BY MARIO LIVIO

Brilliant Blunders: From Darwin to Einstein—Colossal Mistakes by Great Scientists That Changed Our Understanding of Life and the Universe

Is God a Mathematician?

The Equation That Couldn't Be Solved: How Mathematical Genius Discovered the Language of Symmetry

The Golden Ratio: The Story of Phi, the World's Most Astonishing Number

The Accelerating Universe: Infinite Expansion, the Cosmological Constant, and the Beauty of the Cosmos

WHY?

What Makes Us Curious

MARIO LIVIO

Simon & Schuster Paperbacks
NEW YORK LONDON TORONTO SYDNEY NEW DELHI

Simon & Schuster Paperbacks
An Imprint of Simon & Schuster, Inc.
1230 Avenue of the Americas
New York, NY 10020

First Simon & Schuster trade paperback edition May 2018

For information about special discounts for bulk purchases, please contact Simon & Schuster Special Sales at 1-866-506-1949 or business@simonandschuster.com.

The Simon & Schuster Speakers Bureau can bring authors to your live event. For more information or to book an event, contact the Simon & Schuster Speakers Bureau at 1-866-248-3049 or visit our website at www.simonspeakers.com.

Interior design by Paul Dippolito

Manufactured in the United States of America

1 3 5 7 9 10 8 6 4 2

The Library of Congress has cataloged the hardcover edition as follows:

Names: Livio, Mario, 1945– author.
Title: Why? : what makes us curious / by Mario Livio.
Description: New York : Simon & Schuster, 2017. | Includes bibliographical references and index.
Identifiers: LCCN 2016040604| ISBN 9781476792095 | ISBN 1476792097
Subjects: LCSH: Curiosity—History.
Classification: LCC BF323.C8 L58 2017 | DDC 153.3—dc23 LC record available at https://lccn.loc.gov/2016040604

ISBN 978-1-4767-9209-5
ISBN 978-1-4767-9210-1 (pbk)
ISBN 978-1-4767-9212-5 (ebook)

To my mother

Contents

Preface

I HAVE ALWAYS BEEN A VERY CURIOUS PERSON. In addition to my professional interests as an astrophysicist in deciphering the cosmos and various phenomena within it, I have maintained a passion for the visual arts. I have absolutely no artistic talent, but I have amassed a large collection of art books. I am also a science advisor to the Baltimore Symphony Orchestra (yes, there is such a thing), and I have participated in a few of its concerts as a presenter of the links between science and music. Perhaps most exciting from my perspective has been my participation in the creation of the "Hubble Cantata," a contemporary classical music piece by composer Paola Prestini, accompanied by film and virtual reality, all inspired by images taken with the Hubble Space Telescope. In addition, in a regular blog posted on the *Huffington Post*, I tend to informally muse about topics in science and art and the intricate connections between them.

Not surprisingly, therefore, already a long time ago I became intrigued by the questions *What is it that triggers curiosity?* and *What are the underlying mechanisms of curiosity and exploration?* Since this was not my area of expertise, I had to engage in an enormous amount of research, consult with numerous psychologists and neuroscientists, discuss the topic with many scholars from a variety of disciplines, and interview a large number of people whom I believed to be exceptionally curious. As a result, I am deeply indebted to scores of individuals without whom I could not have completed this project. While it would be impractical to attempt to thank them all here, I would like to at least acknowledge a group of people who have both profoundly inspired and greatly informed my writing. I am indebted to Paolo Galluzzi for an illuminating conversation on Leonardo da Vinci and to Jonathan Pevsner for helpful advice on Leonardo and for allowing me to use his vast collection of books and articles on Leonardo. Agata Rutkowska has been a wonderful guide for finding specific Leonardo drawings in the Royal Collection Trust. The Milton S. Eisenhower Library at Johns Hopkins University provided me with hundreds of books on a wide range of relevant disciplines. Jeremy Nathans, Doron Lurie, Garik Israelian, and Ellen-Thérèse Lamm introduced me to people who gave crucial interviews. I am grateful to Joan Feynman, David and Judith Goodstein, and Virginia Trimble for valuable, firsthand information about Richard Feynman.

Jacqueline Gottlieb, Laura Schulz, Elizabeth Bonawitz, Marieke Jepma, Jordan Litman, Paul Silvia, Celeste Kidd, Adrien Baranes, and Elizabeth Spelke provided me with invaluable information, sometimes even before publication, about their research projects in a variety of areas of psychology and neuroscience, all aimed at a better understanding of the nature of curiosity. Any mistakes the book might contain about the interpretation of their results are mine alone. Jonna Kuntsi and Michael Milham clarified for me concepts and potential connections between curiosity and ADHD. Kathryn Asbury discussed with me the implications of various studies involving twins for the nature of curiosity. Suzana Herculano-Houzel explained to me in detail her groundbreaking studies of the constituents of the brain in general and their significance and ramifications for the unique properties of the human brain in particular. Noam Saadon-Grossman helped me navigate the anatomy of the brain. I wish to express my gratitude to Freeman Dyson, Story Musgrave, Noam Chomsky, Marilyn vos Savant, Vik Muniz, Martin Rees, Brian May, Fabiola Gianotti, and Jack

Horner for giving me fabulously interesting and insightful interviews about their own personal curiosity.

Finally, I thank my wonderful agent, Susan Rabiner, for her tireless encouragement and advice. I am grateful to my editor, Bob Bender, for his careful reading of the manuscript and his perceptive and thoughtful comments. General manager Johanna Li, designer Paul Dippolito, copy editor Phil Metcalf, and the entire team at Simon & Schuster again demonstrated their dedication and professionalism in producing this book.

Needless to say, without the patience and continuous support of my wife, Sofie, this book would never have seen the light of day.

WHY?

ONE

Curious

INDEPENDENT OF THEIR LENGTH, SOME STORIES can leave lasting impressions. "The Story of an Hour," a very short tale by the nineteenth-century author Kate Chopin, opens with a rather startling sentence: "Knowing that Mrs. Mallard was afflicted with a heart trouble, great care was taken to break to her as gently as possible the news of her husband's death." Loss of life and human frailty all packed into one punchy line. We then learn that it was the husband's close friend, Richards, who brought the bad news, after having confirmed (by way of a telegram) that Brently Mallard's name was indeed leading the list of those killed in a railroad disaster.

In Chopin's plot, Mrs. Mallard's immediate reaction is a natural one. Upon hearing the sad message from her sister Josephine, she starts weeping straightaway, then retires to her room, asking to be left alone. It is there, however, that something totally unexpected happens. After sitting motionless, sobbing for a while, her gaze apparently fixed on a distant patch of blue sky, Mrs. Mallard starts whispering a surprising word to herself: "Free, free, free!" This is followed by an even more exuberant "Free! Body and soul free!"

When she finally opens the door, yielding to Josephine's worried requests, Mrs. Mallard emerges with "a feverish triumph in her eyes." She starts to calmly descend the steps, clutching to her sister's waist, while her husband's friend Richards awaits them at the bottom of the staircase. That's precisely when someone is heard opening the front door with a latchkey.

Chopin's story contains only eight more lines beyond this point. Could we perhaps stop reading here? Needless to say, even if we wanted to, we probably wouldn't, certainly not without at least knowing who was at the door. As the English essayist Charles Lamb wrote, "Not many sounds in life, and I include all urban and all rural sounds, exceed in interest a knock at the door." That is the power of a story that pulls your attention with such force that you don't even dream of overriding that pull.

The person entering the house is indeed, as you might have guessed, Brently Mallard, who, it turns out, had been so far from the scene of the train accident that he didn't even know it had happened. The vivid description of the emotional roller-coaster ride that the temperamental Mrs. Mallard has had to endure in the span of just one hour turns reading Chopin's drama into a riveting experience.

The last sentence in "The Story of an Hour" is even more unsettling than the first one: "When the doctors came they said she had died of heart disease—of joy that kills." The inner life of Mrs. Mallard remains largely a mystery to us.

Chopin's greatest gift, in my opinion, is her singular ability to generate *curiosity* with almost every single line of prose, even in passages describing situations in which apparently nothing happens. This is the type of curiosity that results from chills running up and down your spine, somewhat similar to the sensation you feel when listening to exceptional pieces of music. Those are subtle, intellectual cliffhangers that constitute a necessary device in any compelling storytelling, lesson at school, stimulating artistic oeuvre, video game, advertising campaign, or even simple conversation that delights rather than bores. Chopin's story inspires what has been dubbed *empathic* curiosity—the standpoint we adopt when we try to understand the desires, emotional experiences, and thoughts of the protagonist and when her or his actions incessantly bother us with the nagging question *Why?*

Another element that Chopin aptly uses is that of surprise. This is a sure stratagem to kindle curiosity through heightened arousal and attention. New York University neuroscientist Joseph LeDoux and his colleagues managed to trace the pathways within our brain that are responsible for the reaction to surprise or fear. When we encounter the unexpected, the brain assumes that some action may have to be taken. This results in a rapid activation of the sympathetic nervous system, with its familiar, associated manifestations: increased heart rate, perspiration, and deep breathing. At the same time, attention is diverted from other, irrelevant stimuli and is focused on the key pressing element under consideration. LeDoux was able to show that in surprise, and in particular in fear response, fast and slow pathways are concurrently activated. The fast track proceeds directly from the thalamus, which is responsible for relaying sensory signals, to the amygdala, an almond-shaped cluster of nuclei that assigns affective significance and directs the emotional response. The slow track involves a lengthy detour between the thalamus and the amygdala that passes through the cerebral cortex, the outer layer of neural tissue that plays a key role in memory and thought. This indirect route allows for a more careful, conscious evaluation of the stimulus and for a thoughtful response.

Several "types" of curiosity—that itch to find out more—exist. British Canadian psychologist Daniel Berlyne charted curiosity along two main dimensions or axes: one extending between perceptual and epistemic curiosity and the other traversing from specific to diversive curiosity. *Perceptual* curiosity is engendered by extreme outliers, by novel, ambiguous, or puzzling stimuli, and it motivates visual inspection—think, for example, of the reaction of Asian children in a remote village seeing a Caucasian for the first time. Perceptual curiosity generally diminishes with continued exposure. Opposite perceptional curiosity in Berlyne's scheme is *epistemic* curiosity, which is the veritable desire for knowledge (the "appetite for knowledge" in the words of philosopher Immanuel Kant). That curiosity has been the main driver of all basic scientific research and of philosophical inquiry, and it probably was the force that propelled all the early spiritual quests. The seventeenth-century philosopher Thomas Hobbes dubbed it "lust of the mind," adding that "by a perseverence of delight in the continual and indefatigable generation of knowledge" it exceeds "the short vehemence of any carnal pleasure" in that indulging in it only leaves you wanting more. Hobbes saw in this "desire to know *why*" (emphasis added) the characteristic distinguishing humankind from all other living creatures. Indeed, as we shall see in chapter 7, it has been the unique ability to ask "Why?" that has brought our species to where we are today. Epistemic curiosity is the curiosity Einstein alluded to when he told one of his biographers, "I have no special talents. I am only passionately curious."

To Berlyne, *specific* curiosity reflects the desire for a particular piece of information, as in attempts to solve a crossword puzzle or to remember the name of the movie you saw last week. Specific curiosity can drive investigators into examining distinct problems in order to understand them better and identify potential solutions. Finally, *diversive* curiosity refers both to the restless desire to explore and to the seeking of novel stimulation to avoid boredom. Today, this type of curiosity might manifest itself in the constant checking for new text messages or emails or in impatience while waiting for a new smartphone model. Sometimes, diversive curiosity can lead to specific curiosity, in that the novelty-seeking behavior may fuel a specific interest.

While Berlyne's distinctions among different types of curiosity have proven to be extremely fruitful in many psychological studies, they should be regarded only as suggestive until a more comprehensive understanding of the mechanisms underlying curiosity emerges. At the same time, a few other types of curiosity have been suggested, such as the empathic curiosity mentioned earlier, which do not neatly fall into Berlyne's categories. There is, for instance, the *morbid* curiosity that results in rubbernecking; it invariably impels drivers to slow down and examine accidents on the highway and prompts people to gather en masse around scenes of violent crimes and building fires. This is the type of curiosity that reputedly generated a huge number of Google hits for the gruesome video showing the beheading of British construction worker Ken Bigley in Iraq in 2004.

In addition to the potentially different kinds, there are also varying levels of intensity that one can associate with assorted genres of curiosity. Sometimes just a snippet of information would suffice to satisfy the curiosity, as in some of the cases of specific curiosity: Who was it who said, "Injustice anywhere is a threat to justice everywhere"? In other instances, curiosity can propel someone into a passionate lifelong journey, as is occasionally the case when epistemic curiosity shepherds scientific inquiry: How did life on Earth emerge and evolve? There are also clear individual differences in curiosity, in terms of the frequency of its occurrence, the intensity level, the amount of time people are prepared to devote to exploration, and in general the openness to and preference for novel experiences. For one person, an old bottle washing ashore on Amrum Island on the German North Sea coast may be just that: a disintegrating symbol of pollution. For another, such a find could trigger an opportunity for a glimpse into an earlier, fascinating world. A message in a bottle found in April 2015 proved to be from sometime between 1904 and 1906—the oldest-known message in a bottle. This was part of an experiment to study ocean currents.

Similarly, Ed Shevlin, a twenty-two-year New York City sanitation worker who collects trash five mornings a week, felt such great enthusiasm for the Gaelic language of Ireland that he enrolled in an NYU master's degree program in Irish American studies.

About two decades ago, a rare astronomical event beautifully illustrated how a few supposedly distinct types of curiosity, such as that evoked by novelty and the one representing the thirst for knowledge, can combine and feed each other to form one irresistible attraction. In March 1993, a previously unknown comet was spotted orbiting the planet Jupiter. The discoverers were veteran comet hunters, husband and wife astronomers Carolyn and Eugene Shoemaker and astronomer David Levy. Since that was the ninth periodic comet identified by this team, the object was named Shoemaker-Levy 9. A detailed analysis of the orbit suggested that the comet had probably been captured by Jupiter's gravity a few decades earlier, and during a catastrophically close approach in 1992, it broke up into pieces due to strong tidal (stretching) forces. Figure 1 presents an image taken by the Hubble Space Telescope in May 1994, showing the resulting two dozen or so fragments as they continued their course along the comet's path like a string of shining pearls.

Excitement in the astronomical world and beyond started to rise when computer simulations indicated that the fragments were likely to collide with Jupiter's atmosphere and to plow into it in July 1994. Such collisions are relatively rare (although one such impact on the Earth some 66 million years ago proved to be extremely un-

Figure 1

fortunate for the dinosaurs) and none had previously been directly witnessed. Astronomers all across the globe were waiting in eager anticipation. Nobody knew, however, if the effects of the impact would actually be visible from Earth or whether the fragments would simply be serenely swallowed by Jupiter's gaseous atmosphere like tiny pebbles by a large, undisturbed pond.

The first icy chunk was expected to hit on the evening of July 16, 1994, and almost every telescope on the ground and in space, including Hubble, was directed at Jupiter. The fact that dramatic astronomical phenomena can seldom be observed in real time (it takes light many years to get to Earth from numerous objects of interest, but only about half an hour from Jupiter) gave this event a "once in a lifetime" feel. Not surprisingly, therefore, a group of scientists, myself included, gathered around a computer screen as the data were about to be transmitted down from the telescope (Figure 2). The question on everybody's mind was: Would we see anything?

If I had to give a title to Figure 2, I know exactly what it would be: *Curiosity!* To feel curiosity's contagious appeal, all you need to

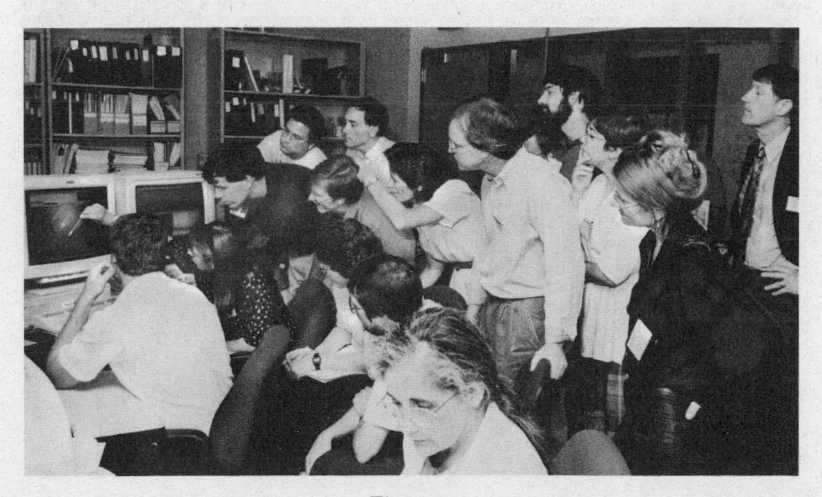

Figure 2

do is examine the posture and facial expressions of the scientists involved. As soon as I saw this photo on the following day, it reminded me of an extraordinary work of art executed almost four hundred years earlier; Rembrandt's *The Anatomy Lesson of Dr. Nicolaes Tulp* (Figure 3). The painting and the photograph are almost identical in how they capture the emotion of impassioned curiosity. What I find especially fascinating is the fact that Rembrandt's focus is neither on the anatomy of the flayed corpse being dissected (though the muscles and tendons are quite accurately depicted), nor even on the identity of the dead man (a young coat thief named Aris Kindt, hanged in 1632), whose face is partially shaded. Rather, Rembrandt was primarily interested in accurately expressing the individual reactions of each of the medical professionals and apprentices attending the lesson. He put curiosity at center stage.

Curiosity's powers extend above and beyond its perceived po-

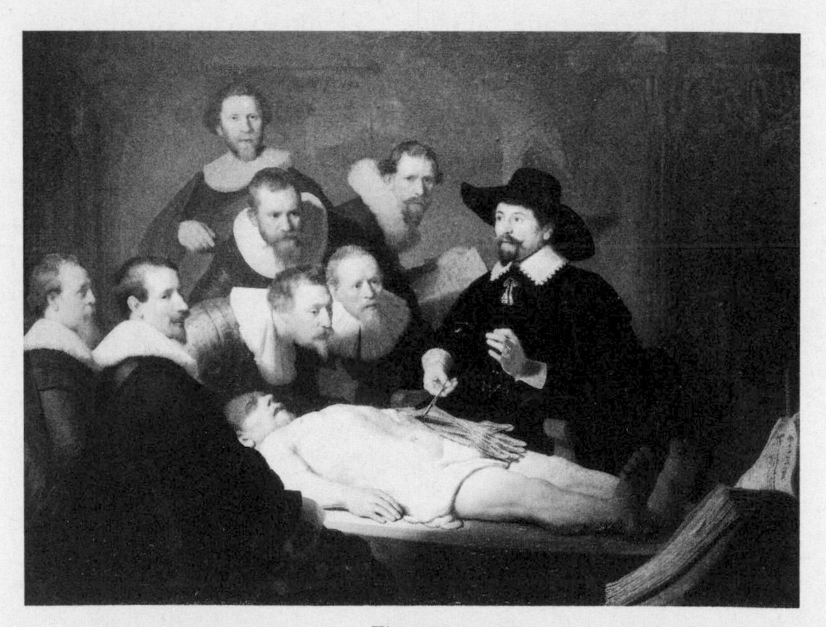

Figure 3

tential contributions to usefulness or benefits. It has shown itself to be an unstoppable drive. The efforts humans have invested, for instance, in exploring and attempting to decipher the world around them, have always far exceeded those needed for mere survival. It seems that we are an endlessly curious species, some of us even compulsively so. University of Southern California neuroscientist Irving Biederman says human beings are designed to be "infovores," creatures that devour information. How else would you explain the risks people sometimes take to scratch that curiosity itch? The great Roman orator and philosopher Cicero interpreted Ulysses's sailing past the island of the Sirens as an effort to resist epistemic curiosity's lure: "It was not the sweetness of their voices, nor the novelty and diversity of their songs, but their professions of knowledge that used to attract the passing voyagers; it was the passion for learning that kept men rooted to the Sirens' rocky shores."

French philosopher Michel Foucault beautifully describes a few of curiosity's inherent characteristics: "Curiosity evokes 'care'; it evokes the care one takes of what exists and what might exist; a sharpened sense of reality, but one that is never immobilized before it; a readiness to find what surrounds us strange and odd; a certain determination to throw off familiar ways of thought and to look at the same things in a different way; a passion for seizing what is happening now and what is disappearing; a lack of respect for the traditional hierarchies of what is important and fundamental."

As we shall see, modern research suggests that curiosity may be essential for the proper development of perceptual and cognitive skills in early childhood. There is also little doubt that curiosity remains a powerful force for intellectual and creative expression later in life. Does this mean that curiosity is a straightforward product of natural selection? If it is, why do even seemingly trivial matters sometimes make us vehemently curious? Why do we occasionally strain to decipher the hisses of a conversation at the table next to us in a restaurant? Why do we find it harder not to listen to someone talking on the phone (when we hear only half of the conversation) than to listen to two people having a face-to-face exchange? Is curiosity entirely innate, or do we learn to become curious? Conversely, do adults lose their childhood curiosity? Has curiosity evolved during the 3.2 million years that separate Lucy—the transitional, nearly human creature whose bones were found in Ethiopia—from the *Homo sapiens*, modern humans? Which psychological processes and which structures within our brains are involved in being curious? Is there a theoretical model of curiosity? Do some neurodevelopmental disorders such as ADHD represent curiosity "on steroids" or curiosity spinning its gears?

Before seriously delving into the scientific research on curiosity, I decided (out of my own personal curiosity) to take a brief detour to closely examine two individuals who, in my view, represent two of the most curious minds to have ever existed. I believe that few would disagree with this characterization of Leonardo da Vinci and the physicist Richard Feynman. Leonardo's boundless interests spanned such broad swaths of art, science, and technology that he remains to this day the quintessential Renaissance man. Art historian Kenneth Clark appropriately called him "the most relentlessly curious man in history." Feynman's genius and achievements in numerous branches of physics are legendary, but he also pursued fascinations with biology, painting, safecracking, bongo playing, attractive women, and studying Mayan hieroglyphs. He became known to the general public as a member of the panel that investigated the space shuttle *Challenger* disaster and through his best-selling books, which are chock-full of personal anecdotes. When asked to identify what he thought was the key motivator for scientific discovery, Feynman replied, "It has to do with curiosity. It has to do with wondering what makes something do something." He was echoing the sentiments of the sixteenth-century French philosopher Michel de Montaigne, who urged his readers to probe the mystery of everyday things. As we shall see in chapter 5, experiments with young children have demonstrated that their curiosity is often triggered by the desire to understand cause and effect in their immediate surroundings.

I don't expect that even a careful inspection of the personalities of Leonardo and Feynman will necessarily reveal any deep insights into the nature of curiosity. Numerous previous attempts to uncover common features in many historical figures of genius, for instance, have exposed only a perplexing diversity with respect to the backgrounds and psychological characteristics of these individuals. Take the scientific giants Isaac Newton and Charles Darwin. Newton was distinguished by his unparalleled mathematical ability, while Darwin was, by his own admission, rather weak in mathematics. Even within classes of masterminds in a given scientific discipline, there appears to be an ambiguous array of qualities. Physicist Enrico Fermi solved very difficult problems at age seventeen, while Einstein was, relatively speaking, a late bloomer. This is not to say that *all* efforts to identify a few shared characteristics are doomed to fail. In the area of prodigious creativity, for example, University of Chicago psychologist Mihaly Csikszentmihalyi has been able to unearth a few tendencies that appear to be associated with most unusually creative persons (those are briefly described at the end of Chapter 2). I therefore thought it a worthwhile exercise at least to explore whether there was anything in the fascinating personalities of Leonardo and Feynman that could provide a clue about the source of their truly insatiable curiosity. The key point for me was the fact that irrespective of whether Leonardo and Feynman had anything in common other than their curiosity, they both stood so high above their respective surroundings in terms of their spirit of inquiry that any stab at viewing things from their perspective was bound to be stimulating. I start with Leonardo, who once elegantly expressed his own passion for comprehension by saying, "Nothing can be loved or hated unless it is first understood."

By the way, in case you are curious to know whether we actually saw anything when the first fragment of Comet Shoemaker-Levy 9 hit Jupiter's atmosphere—we did! At first there was a point of light above Jupiter's edge. As the fragment penetrated the atmosphere, it produced an explosion that resulted in a mushroom cloud similar to that created by a nuclear weapon. All the fragments left visible "scars" (areas with sulfur-bearing compounds) on Jupiter's surface (Figure 4). Those smudges lasted for months until they were smeared out by streams and turbulence within Jupiter's atmosphere, and the debris diffused down to lower altitudes.

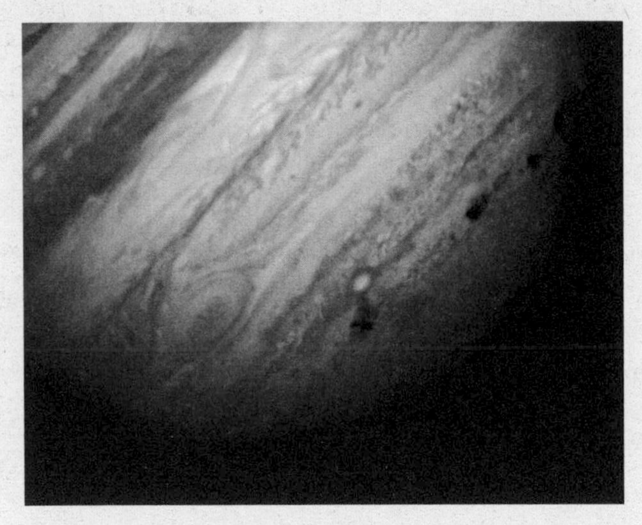

Figure 4

TWO

Curiouser

THE IMAGE WE HAVE TODAY OF LEONARDO DA Vinci was perhaps best encapsulated by two short sentences of Giorgio Vasari, author of the celebrated *Lives of the Most Eminent Painters, Sculptors and Architects*, who was just eight years old when Leonardo died. Vasari wrote in admiration, "Besides a beauty of body never sufficiently extolled, there was an infinite grace in all his actions; and so great was his genius, and such its growth, that to whatever difficulties he turned his mind, he solved them with ease." I would have made only one small revision to this description, to read "so great were his genius and *curiosity*, and such their growth."

When he came to elaborate on those splendid attributes, Vasari emphasized Leonardo's great capacity to rapidly learn new subjects in an astonishing variety of disciplines: "In arithmetic, during the few months that he studied it, he made so much progress, that, by continually suggesting doubts and difficulties to the master who was teaching him, he would often bewilder him. He gave some little attention to music, and quickly resolved to learn to play the lyre, as one who had by nature a spirit most lofty and full of refinement: wherefore he sang divinely to that instrument, improvising upon it." In light of these effusive accolades, it may come as a surprise that more recent studies have revealed that Leonardo's notes in mathematics actually contain some embarrassing mistakes and oversights, for example in the extraction of roots. Similarly, Leonardo could not read Greek, and he read even Latin with difficulty, usually assisted by knowledgeable friends. On the face of it, these two traits, an incredible ability to acquire new knowledge coupled to bewildering gaps in basic education, appear to be in serious conflict. Two facts, however, provide at least a starting point for an explanation. First, Leonardo's early education at Vinci was rather rudimentary, and when he apprenticed in Master Andrea del Verrocchio's atelier in Florence, he trained to be an artist, not a scientist, mathematician, or engineer. Accordingly, he studied basic reading and writing, supplemented by techniques in painting, sculpting, and some practical rules of geometry and mechanics and those practices needed for metalworking. No one could have predicted that from such inauspicious beginnings Leonardo would rise to become the symbol of the Renaissance ideal of the universal man. All of his eventual, seemingly all-encompassing erudition was self-taught or gained from ceaseless experiments and observations much later in life. In fact, as a result of the void in his command of the classics, the humanistic scholars of Leonardo's time condescendingly repeated his own self-description as "a man without letters" or "not well read." Leonardo himself was quick to add, however, "Those who study the ancients and not the works of Nature are stepsons and not sons of Nature, the mother of all good authors." Defying the critics he continued, "Though I may not like them, be able to quote other authors, I shall rely on that which is much greater and more worthy—on experience, the mistress of their masters." Leonardo was without a doubt the archetypal "disciple of experience."

Vasari also provides us with a second clue that could potentially demystify the contradictory facets in Leonardo's education: "For he set himself to learn many things, and then, after having begun them, abandoned them." In other words, Leonardo did not persist in some of his studies. This, however, introduces a new puzzle: Why would Leonardo abandon topics in which he had initially shown great interest? This is an important question to which we shall return, since it may provide some insights into the workings of Leonardo's curiosity-driven mind.

To simply state that Leonardo was a curious person would be the understatement of the millennium. Suffice it to note that even a partial inventory of his library in 1503–4 contains no fewer than 116 books covering a staggering breadth of topics. Those range from anatomy, medicine, and natural history, through arithmetic, geometry, geography, and astronomy, all the way to philosophy, languages, literary works, and even religious treatises. And this was the library of a man who, by all accounts, much preferred experimentation to reading—so much so, in fact, that historian of science and Leonardo scholar Giorgio de Santillana entitled one of his lectures "Leonardo and Those He Did Not Read."

One of the most intriguing aspects of Leonardo's personality is the apparent conflict between his compassionate aesthetic sensibility and his unemotional, superhumanly sharp eye when it came to analyzing the secrets of nature. The physician and historian Paolo Giovio provided us in 1527 (only eight years after Leonardo's death) with an introduction to Leonardo's unique view of what he regarded as the inescapable links between science and art. Giovio wrote, "Leonardo da Vinci . . . added great lustre to the art of painting, denying that this could be properly carried out by those who had not attained the noble sciences and liberal arts necessary to the disciples of painting." To illustrate Leonardo's distinctive approach, Giovio describes in succession a few of the many scientific activities undertaken by the master in conjunction with painting: "The science of optics was to him of paramount importance. . . . He dissected the bodies of criminals in the medical schools . . . in order that the variations of the joints of the limbs flexed by the actions of the nerves of the vertebrae should be painted according to the laws of nature."

Giovio's report correctly conveys the important fact that in his early work Leonardo used nature as a servant of his art: he examined the natural world to make his artistic representation as accurate as possible. Later in life, however, art became the obsequious assistant of his scientific investigations: he used his singular artistic ability to depict natural phenomena and to attempt to ascertain their causes.

As early as two decades before Vasari, Giovio too commented on Leonardo's apparent inability to complete assignments, or his lack of interest in finalizing some of his projects: "But while he was thus spending his time in the close research of subordinate branches of his art he carried only very few works to completion." Even during his lifetime, Leonardo's tendency to leave undertakings unfinished was legendary. When Pope Leo X heard that Leonardo was fussing over various recipes for varnishes instead of actually painting, he apparently complained, "Alas! This man will never do anything, for he begins by thinking about the end before the beginning of his work."

To Leonardo, ostensibly, each painting was also a scientific experiment, both in terms of correctly presenting the subject matter being portrayed and with regard to the execution of the painting itself. It was an exercise in curiosity as well: "Study the science of art; Study the art of science; Learn to see," he said. Concerning the physical implementation of painting techniques, some of his paintings, *The Last Supper* (Figure 5) for instance, failed to last—the paint was probably flaking off the wall even in Leonardo's lifetime. From a different standpoint, however, *The Last Supper* is a resounding success and an extraordinary masterpiece. It represents a brilliant study in perspective and in the effective use of light and shadow. One can even perceive in the spreading of the emotional wave created by Christ's words "One of you shall betray me" the lessons Leonardo had learned from his observations of the propagation of waves in water.

Figure 5

Here, however, lies another contradiction. The same person who was capable of so delicately capturing the most subtle human moods and emotions (also seen in *The Virgin and Child with St. Anne* and the celebrated *Mona Lisa*) revealed almost nothing of his personal feelings in his extensive writings. If Leonardo was as curious about his own internal world as he was about the external one, he chose not to let us in.

Amazed and Curious

Many excellent studies have attempted to use Leonardo's numerous notebooks, detailed comments, and elaborate drawings to assess his actual achievements and the degree to which he produced genuinely new discoveries in science and technology. Others have sought to critically evaluate the originality of his contributions, given the existing knowledge at the time. I am interested in different questions that are equally enticing: What was it that made

Leonardo curious, and why? What did he do to satisfy his curiosity? At what point, if any, did he actually lose interest in a particular topic? Rather than being concerned with Leonardo's successes and failures in his scientific endeavors and his artistic and engineering projects, or with the extent to which he did or did not influence scientific progress or the course of art history, I am curious about what caught his fancy, what motivated him, and how he responded to such stimuli.

Leonardo's personal notebooks are an excellent point of departure for addressing these questions, for the following main reasons. First, the extant 6,500 pages of notes and drawings probably represent only part of his output, estimated by some researchers to be 15,000 pages. Since Leonardo began to keep notebooks only around the age of thirty-five, he must have filled on average, about a page and a half every day for three decades! It would seem that loading pages with painstaking drawings and sophisticated notes describing his ideas, interests, and contemplations (mostly written with his left hand, from right to left, and in mirror image) was one of Leonardo's most cherished occupations. Amazingly, the existing body of Leonardo's drawings alone is about four times that of even the most productive sixteenth-century draftsman. Second, aside from this apparent obsession with analyzing and documenting every rational thought, the actual content of the notebooks covers such topics as anatomy, vision and optics, astronomy, botany, geology, physical geography, the flight of birds, movement and weight, the properties and motion of water, and a confounding variety of imaginative inventions for both peaceful purposes and warfare. Finally, couple the vast scientific and technological substance of the notebooks to the reality that Leonardo used those very same pages for incessant comments on artistic matters, such as color, light and shade, perspective, precepts of the painter, and sculpture and architecture. The picture that emerges is as clear, and at the same time as enigmatic, as some of the elements in Leonardo's paintings themselves.

Leonardo was curious about almost *everything* in the complex world surrounding him, and his compulsive note taking and drawing represented his idiosyncratic attempt to make sense of it all. To be sure, he was never particularly interested in history, theology, economics, or politics (probably wisely so, since he lived during the period in which the notoriously conniving and cruel Borgias were in power). He was nevertheless trying to "read" and decipher what Galileo Galilei would call more than a century later the "book of Nature." However, Leonardo's book of nature was an even thicker volume than Galileo's, since it included such complex topics as anatomy and botany, subjects in which Galileo did not show great interest. On the whole, the vast majority of the entries in Leonardo's notebooks were not intended to be blueprints, preparatory sketches, or engineering plans supposed to culminate in physical implementations of specific ventures. Rather, they were Leonardo's curiosity incarnate. In his own words, "Nature is full of infinite causes which were never set forth in experience. . . . The natural desire of good men is knowledge." Leonardo was anticipating here what psychiatrist Herman Nunberg would say almost five centuries later: "Through the gratification of curiosity one acquires a certain stock of knowledge, which may again lead to new problems and the formulation of new questions. Curiosity may therefore also be called an *urge for knowledge*."

The notebooks also graphically demonstrate the powerful interdependence of science, technology, and art in Leonardo's mind. The phrase "a picture is worth a thousand words" is thought to have originated from a newspaper article in 1911, yet Leonardo clearly expressed this very sentiment four centuries earlier: "You who think to reveal the figure of man in words . . . banish the idea from you, for the more minute your description, the more you will confuse the mind of the reader, and the more you will lead him away from the knowledge of the thing described. It is necessary, therefore, for you to represent and describe."

The drawings, however, do much more than merely illustrate themes that are difficult to describe in words. They sometimes allow us literally to follow the serpentine meanderings of Leonardo's curiosity. A wonderful example is provided by a work in the Royal Collection (Figure 6). Carlo Pedretti, a Leonardo scholar, remarked that this single sheet may offer "a complete synthesis of his [Leonardo's] scientific curiosity and his artistic versatility."

At first blush, the page appears to contain nothing more than a series of unrelated scribbles: various geometrical constructions involving circles and curves, clouds, weeds climbing a lily, a screw press, a clothed old man, waves in a pond, branches of a tree. Yet a closer examination reveals that almost every doodle, from the bil-

Figure 6

lowing clouds to the man's curly hair, involves geometrical curves, curved surfaces, or the phenomenon of branching. We can therefore speculate that once Leonardo started to contemplate a particular phenomenon, such as the propagation of waves in a pool, his visually inspired mind immediately translated the problem into that of a geometrical shape. Concomitantly, his wandering curiosity guided him to an entire menagerie of other natural phenomena or human-made devices in which similar curves or geometrical structures appear. For example, when magnified, the drawing shows the branches of the tree metamorphosing into a network of veins, seen through the old man's cape.

That was not the only time that Leonardo examined branching systems. He noticed those structures across a wide range of different disciplines, from the tributaries of rivers, through the stems of plants, to the blood vessels in the human body. The culmination of the dizzying mental journey that led to the creation of Figure 6 was the abstraction of a common feature from a collection of seemingly disparate observations. In Leonardo's own words, "Painting compels the mind of the painter to transform itself into the very mind of Nature to become an interpreter between Nature and the art. It explains the causes of Nature's manifestations as compelled by its laws."

Given the scientific backdrop against which Leonardo was working, this last statement is truly remarkable. He is asserting that Nature is governed by certain laws! This is about a century before Galileo articulated his law of inertia and almost two centuries before Newton formulated his laws of motion and gravitation. Was Leonardo also sufficiently curious to wonder what those laws might be? You can bet he was. Unfortunately, the scientific tradition of his time did not yet include the statement of a coherent hypothesis and the testing of that hypothesis through a series of carefully constructed experiments or observations. Instead, Leonardo tended to simply list all the questions he could think of, probably in the order in which those popped into his relentlessly curious mind, and then to address only a few of those through more careful inspections. Sometimes, however, what he discovered was a fusion of his artistic and scientific visions. For instance, his drawings of water flows often resemble braids of hair, and the wavy hair in his painting of Ginevra de' Benci (Figure 7) looks like turbulent water. Still, from a multitude of diverse studies, Leonardo did emerge with two major realizations. First, he concluded that repeated, quantitative experiments and observations were absolutely crucial for the incontrovertible detection of the patterns associated with natural phenomena. In his words, "This experiment should be made many times so that no accident may occur to hinder or falsify this proof, for the experiment might be false whether it de-

Figure 7

ceived the investigator or not." This may partially explain the fact that Leonardo's notebooks contain many repetitions, even though his quantitative measurements are approximate at best. His second remarkable deduction was that the human mind could gain access to the governing laws of nature through the language of mathematics. Accordingly, much of Leonardo's work during the last two decades of his life was devoted to the pursuit of general geometrical laws that would apply to phenomena ranging from the currents in rivers to light and shade and the intricacies of human anatomy.

Following in Plato's and the Neoplatonists' footsteps, geometry became Leonardo's guiding light on the pathway connecting the human observer to the explanations and interpretations of the universe, even if that connection was more a matter of belief than relying on a solid empirical foundation. First, there was the geometry associated with the process of vision, then the geometrical rules or laws that the natural world was supposed to obey, and finally the nature of the mathematical language itself, which to Leonardo was the basic Euclidean geometry we learn at school. Concerning the propagation of light, for instance, Leonardo drew a series of triangles ("pyramids" in his terminology) and concluded (incorrectly in quantitative terms) that the light intensity decreases proportionally to the inverse of the distance from the source, that is, that a source twice as distant would appear half as bright. In reality, the brightness decreases with an inverse square law: at twice the distance a light source appears four times dimmer, at three times the distance nine times dimmer, and so on. He applied similar laws to what he defined as the four "powers" of nature: "movement, force, weight, and percussion."

For branching systems, such as trees, Leonardo introduced an original law, according to which the sum total of the cross-sectional areas at each level must be equal. For instance, he inferred that the sum of the areas of the cross sections of the smallest twigs in the periphery must be equal to the cross-sectional area of the tree trunk. While the idea underlying this pronouncement was ingenious and correct (Leonardo deduced that what flows in must flow out), he neglected the fact that the speed of the flow could vary along the way, and consequently his law was not accurate. From our perspective, however, the important point is not whether Leonardo's rules were correct or whether he knew enough mathematics to even attempt to formulate precise laws. The crucial element is the fact that he used a geometrical representation of rules at all. Moreover, he contended that "there is no certainty where one cannot apply any of the mathematical sciences or any of those which are connected to the mathematical sciences." This exceptional insight is comparable to Galileo's famous dictum "We cannot understand it [the universe] if we do not first learn the language and grasp the characters in which it is written. It is written in the language of mathematics, and the characters are triangles, circles and other geometrical figures." But Galileo was a mathematician. Astonishingly, Leonardo, who was rather weak in mathematics, except perhaps in some aspects of curvilinear geometry (and a few elements he learned from his mathematician friend Luca Pacioli), already believed that the only way to understand the universe with some certainty was through mathematics. Consequently, he was bold enough to write, "No man who is not a mathematician should read the elements of my work"—a phrase very reminiscent of the legendary inscription that supposedly hung above the door to Plato's Academy: "Let no one destitute of geometry enter."

One of Leonardo's key comprehensions was that, irrespective of what the laws were, they were in some sense *universal.* That is, the same laws applied to all the "powers," whether those powers acted in the macrocosm of the world at large, in the microcosm represented by the human body, or in the workings of human-made machines. He wrote, "Proportion is not only found in numbers and measurements but also in sounds, weights, times, spaces, and whatsoever powers there be." Similarly, in his correct anticipation of Newton's third law of motion (that any reaction is equal in strength but opposite in direction to the action), Leonardo wrote, "An object offers as much resistance to the air as the air does to the object." This was immediately followed by "And it is the same with water."

Eventually, as part of his aspiration to find general laws or broad-ranging distinctive features and to apply those to specific situations, Leonardo turned his attention to the human body. In that arena, as University of Toronto professor of anatomy James Playfair McMurrich writes, "If . . . the impulse to the new movement in anatomy came from the artists, Leonardo may well be recognized as its originator and Vesalius [the anatomist Andreas Vesalius, who was born five years before Leonardo's death] as its great protagonist."

All Thy Heart Lies Open unto Me

Perhaps the best example of Leonardo's curiosity in action is provided by his unflagging investigations into the operation of the human heart. That mysterious constant beat in the chest has fascinated humans since antiquity. Yet even though partially correct ideas identifying the heart as a pump circulating the blood appeared in China as early as the second century BCE, for a long time those concepts did not penetrate into the prevailing Western theory. The latter was dominated until the sixteenth century by the teachings of the second-century CE Greek physician Galen of Pergamum. Galen concluded that the heart was not a pump, but rather acted as the body's vitalizing fireplace, generating internal heat. Ironically, even though Galen himself was an extremely curious person who based his anatomical observations on actual dissections of monkeys, pigs, and dogs, most of his followers blindly accepted his conclusions for more than a millennium. Just as Aristotelian views reigned supreme in the physical sciences, and the Ptolemaic geocentric model of the solar system remained unchallenged, so in anatomy Galen's theories were considered sacred. It is as if curiosity had frozen solid during the Middle Ages. Leonardo, on the other hand, took to heart Galen's advice: "We must be daring and search after Truth; even if we do not succeed in finding her, we shall at least be closer than we are at present."

According to Galen, when the heart dilates, it draws air from the lungs. This air passes into the left ventricle, where it mixes with blood, producing "vital spirits" by virtue of the "innate heat." When the heart contracts, the blood and vital spirits exit through the arteries, reaching and "vivifying" all the tissues.

Leonardo's interest in the heart was so profound that he devoted more space in his notebooks to the heart than to any other organ (Figure 8 shows two of his drawings of the heart, probably of an ox). Unfortunately, even he couldn't totally rid himself of Galen, with whose ideas he had become acquainted mostly through the works of the tenth-century Persian polymath Avicenna (Latinate form of Ibn-Sina) and the thirteenth-century Italian physician Mondino de Luzzi.

It is somewhat regrettable that Leonardo used Avicenna's *The Canon of Medicine* and de Luzzi's *The Anatomy of the Human Body* as the starting point for his own explorations, since in a few cases even this partial adherence to the older texts led him astray, or at least to unnecessary errors. Nevertheless, through his own meticulous investigations and experiments, Leonardo did manage to dispose of most of Galen's obscure concepts, such as "innate heat" and the mysterious "natural and animal spirits," and he replaced those with physical phenomena associated with standard fluid motion. To Leonardo, "the heart in itself is not the beginning of life; but it is

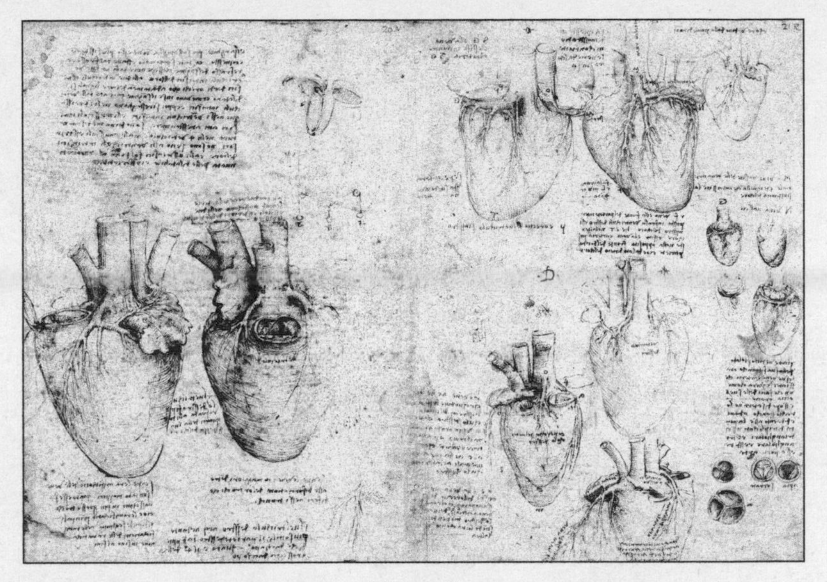

Figure 8

a vessel made of thick muscle, vivified and nourished by the artery and vein, as are the other muscles."

From this simple but fundamental appreciation, he went on to discover parts of the heart not even mentioned by Galen, most notably the atria. Leonardo correctly identified these as contracting chambers that push blood into the ventricles. At an even more basic level in terms of the underlying physical processes involved, he suggested that the heat, which he regarded as a signature of life, is generated by friction with the ebb and flow of the blood. He then used this idea to explain the association of fever with a more rapid pulse: "The more rapidly the heart moves the more the heat is increased, as the pulse of the febrile, moved by the beating of the heart, teaches us."

In the spirit of exploration, Leonardo employed a combination of ingenious experiments and fastidious observations to unscramble the functions of various parts of the heart. In some of his tests, he creatively represented the aorta by a glass model and the ventricle by a flexible bag. On the observational side, he used analogues of blood flow by following the motion of seeds in a fluid, in the same way he had previously investigated the flow of water in rivers.

The chief obstacle that ultimately prevented Leonardo from discovering and understanding the entire concept and mechanism of blood circulation was probably the fact that he had never witnessed a dissection of the chest of a live human being. Consequently, he missed the opportunity to see with his own eyes what he surely would have regarded as a wonderful machine—the human heart—while it was still beating. The comprehensive understanding of the circulatory system was left for the English physician William Harvey, more than a century later. What Leonardo did accomplish through his tenacious scrutiny was still quite remarkable. Single-handedly, he removed almost all of Galen's unnatural elements from the description of the life processes, and he placed life itself squarely within the realm of general physical laws. His clear and prescient judgment marked the dawn of the scientific awakening that was about to follow: "Why Nature cannot give the power of movement to animals without mechanical instruments, as is shown by me in this book on the works of movement which nature has created in animals. And for this reason I have drawn up the rules of the four powers of nature."

Simply put, Leonardo replaced the mystic black bile, faculties, and spirits that permeated the writings of Galen, Avicenna, de Luzzi, and others, with his physical powers of movement, weight, force, and percussion—the building blocks of mechanics. He further used these mechanical concepts to demystify a whole host of physiological processes. For instance, he correctly described the pulse thus: "Expansion occurs when they [the vessels] receive the excessive quantity of blood, and contraction is due to the departure of the excess of blood they have received."

There is no doubt that in spite of the fact that many of his methods were nonscientific by modern standards, in his striving to explain phenomena through physical rather than supernatural effects, Leonardo represented the burgeoning of modern thinking about the true nature of scientific research. His was the type of observation-based, empirical exploration that eventually led to such curious and towering scientists as Galileo, Newton, Michael Faraday, and Darwin, and empiricist philosophers such as John Locke, who argued that knowledge is achieved through the perception of the senses and rational contemplation rather than being planted in the mind by a divine power.

I Have Seen a Curious Child

What was it, then, that distinguished Leonardo from his predecessor anatomists, hydraulicists, botanists, and technologists? And why did he, trained as an artist, succeed in producing scientific and technological discoveries that, even if occasionally wrong, were at times far ahead of those of even his professional contemporaries? After all, the opportunities he had to become involved in, say, anatomical studies, were available to any other scientist and artist of his time. The answer to these questions is actually so simple that it sounds almost banal: Leonardo had an unquenchable curiosity which he attempted to satisfy directly through his own observations rather than by relying on statements by figures of authority. It was neither the result of a particular investigation nor even the method used in any specific inquiry that significantly distinguished Leonardo from his contemporaries. It was the fact that he considered almost every natural phenomenon interesting and worthy of study.

What if his observations did not agree with prevailing wisdom? Leonardo had an unambiguous answer: in that case it was the theory that needed to be revised or altogether rejected. In his words, "Wrongly do men blame innocent experience, accusing her of deceit and false results. . . . Experience is not at fault, it is only our judgment that is in error in promising itself from experience things that are not in her power!"

Take the field of anatomy as an example. Whereas to many medieval anatomists, dissection served merely as a demonstration of Avicenna's teachings, Leonardo dissected to explore and prove things to himself. Similarly, in mechanics, while Leonardo's earliest writings did consider some contemporary ideas for perpetual motion machines, by 1494, following results from his own experiments, he had convinced himself that at least some designs would not work: "Oh! Speculators about perpetual motion, how many vain chimeras have you created in the like quest. Go and take your place with the seekers after gold!"

As I have already noted, there are a few personality characteristics of Leonardo that deserve special attention. First, there was the apparent contradiction between his being rather reclusive and his obsessive documentation of every conception, presumably at least partially for others to eventually read. One of the speculations about his mirror writing is that he was trying to make it harder for people to read his notes, but as we shall soon see, that may not have been the case.

Second, there was the discrepancy between Leonardo the cold, seemingly emotionless analyzer of the natural world and Leonardo the tender, almost romantic painter of exquisitely nuanced human feelings. In his entire oeuvre, only once did he truly reveal an aspect of his emotional side in writing (as he regularly did in painting). In his description of a journey he took to the mountains he wrote:

> Having wandered some distance among gloomy rock, I came to the entrance of a great cavern, in front of which I stood some time, astonished and unaware of such a thing. Bending my

> back into an arch I rested my left hand on my knee and held my right hand over my down-cast and contracted eye brows; often bending first one way and then the other, to see whether I could discover anything inside, and this being forbidden by the deep darkness within, and after having remained there some time, two contrary emotions arose in me, fear and desire—fear of the threatening dark cavern, desire to see whether there were any marvelous things within it.

As we shall see in chapter 4, unknowingly Leonardo captured here one of the suggested characteristics of curiosity: an ambivalent combination of excitement and apprehension. Up to a point, uncertainty about a topic enhances curiosity. After that point, however, the uncertainty becomes so overwhelming that it can produce discomfort, or even fear.

Leonardo's passion for discovering new things within the yet unexplored parts of the world is also reminiscent of another staggeringly brilliant but otherwise socially challenged individual, Isaac Newton. Shortly before his death, Newton said, "I do not know what I may appear to the world; but to myself I seem to have been only like a boy playing on the seashore, and diverting myself in now and then finding a smoother pebble or a prettier shell than ordinary, whilst the great ocean of truth lay all undiscovered before me." Einstein, another famously curious person, talked about "this huge world, which exists independently of us human beings and which stands before us like a great, eternal riddle, at least partially accessible to our inspection and thinking."

Third, there was the issue of Leonardo's having been extremely keen on acquiring new projects, either for investigation or for execution, but rarely completing them. How can we explain these opposing traits in Leonardo's personality, and are they somehow related to his ravenous curiosity?

Interestingly, it is precisely this unusual ability to move from one extreme to the other, to express both ends of the continuum of a trait, that Csikszentmihalyi identifies as the main quality that distinguishes creative personalities from others. He terms this peculiarity "complexity." In his words, "Instead of being an 'individual,' each of them [creative people] is a 'multitude.'" As an illustration of what he means by "complexity," Csikszentmihalyi lists pairs of apparently opposing characteristics that creative people paradoxically seem to exhibit. These include, for instance, great physical activity coupled with frequent periods of quiet and rest; responsibility and irresponsibility; the ability to alternate between imagination and fantasy on one hand, and a rooted sense of reality on the other; opposite tendencies on the range between extroversion and introversion; and even "psychological androgyny," an uncommon combination of mannerisms typically classified as "feminine" with others classified as "masculine."

An examination of this list shows that it fits Leonardo perfectly. Concerning the last oddity, Leonardo was thought by many researchers, including Sigmund Freud, to have been homosexual, although perhaps latent. He also appeared to have undergone an extreme transition, from a strong sexual ardor as an infant to a cool asexuality as an adult. The degree to which he matches the description of a complex personality is hardly surprising, since he was clearly a supremely creative person. Does this mean that being curious and being creative are one and the same thing? Even though people often confuse the two characteristics, they are not identical. A creative person is someone whose ideas or activities significantly change an existing cultural domain or create a new one. Simply being curious is not a sufficient condition for creativity. Curiosity does appear, however, to be a *necessary* condition for creativity. Indeed, Csikszentmihalyi found that practically every creative person that he had ever interviewed or examined exhibited a more than usually keen curiosity.

An amusing anecdote involving Darwin epitomizes the power of curiosity in creative people. When Darwin arrived at Cambridge in 1828, he became an avid collector of beetles. Once, after stripping the bark from a dead tree, he found two ground beetles and caught one in each hand. At that point, he caught sight of a rare crucifix ground beetle. Not wanting to lose any of them, he popped one beetle in his mouth to free a hand for the rarer species. That particular adventure did not end well. The beetle in Darwin's mouth released an irritating chemical and he was forced to spit it out, apparently losing all three beetles in the process. The disappointing result notwithstanding, the story does demonstrate curiosity's irresistible appeal.

There is another intriguing aspect of Leonardo's personality. Consider the following list of "symptoms" a person may exhibit:

- Be easily distracted, forget things, and frequently switch from one activity to another
- Have difficulty focusing on one thing
- Become bored with a task after only a few minutes, unless doing something enjoyable
- Have difficulty focusing attention on organizing and completing a task or learning something new
- Have trouble completing or turning in assignments
- Dash around, touching or "playing" with everything in sight
- Be constantly in motion

One could argue that to some extent Leonardo exhibited most if not all of these symptoms. Yet this is a partial list of the symptoms used to diagnose people affected by Attention Deficit Hyperactivity Disorder (ADHD). Could it be that Leonardo's shifting interests and difficulties in completing projects were a manifestation of the fact that he suffered from ADHD? Or is this just another case of cyberchondria—a diagnosis based on apparent symptoms that is induced by an internet search? More important from our perspective, is there any known or even suspected connection between ADHD and prodigious curiosity?

We cannot expect to achieve a reliable diagnosis for a person who has been dead for almost five centuries, and I do not pretend to be a psychobiographer. Still, I was sufficiently intrigued by this last question to consult with a few experts. In particular, I wondered if a person with ADHD could concentrate for relatively long periods of time on a particular subject, as Leonardo obviously had done.

"Absolutely," Jonna Kuntsi, an ADHD researcher at London's King's College, told me. "Adults with ADHD can focus their attention when they are really interested in something. In fact, even children with ADHD were found to concentrate very well when engaged in computer games that attracted them." Kuntsi pointed out that some people with ADHD were able to put it to good use. An example is British Olympic medal-winning gymnast Louis Smith, who turned ADHD and a strict training regimen into a winning combination.

Michael Milham, a neuroscientist at New York's Child Mind Institute, agreed with Kuntsi. "ADHD could lead someone with high intelligence to thinking 'outside the box,'" he said.

Is there any known correlation between curiosity and ADHD? Kuntsi directed me to a series of studies that have demonstrated a relationship between hyperactivity-impulsivity and the temperament characteristic of novelty seeking—one of the key manifestations of diversive curiosity. In other words, distractibility can be regarded as an acute overexpression of curiosity. What could be the theoretical-physiological basis for such a connection? Both Kuntsi and Milham explained that there is considerable research pointing to ADHD being most likely related to the level of the neurotransmitter dopamine, a chemical that transmits signals between nerve cells, which plays a major role in the brain's reward system. Therefore, if there really is such a link, it would suggest an association between curiosity and reward. Is there research supporting such a coupling? There definitely is, as we will explore in chapters 5 and 6, when I discuss in detail the processes in the brain that are related both to the arousal of curiosity and to its satisfaction.

Returning to Leonardo and his interests, it seems that he stuck with a topic for just as long as he was still curious about it, but no longer. Once he had assuaged his curiosity about a certain project, he saw no point in continuing to work on it. Did he suffer from ADHD? We shall probably never know, but neither Kuntsi nor Milham laughed at the idea. As Bradley Collins wrote in his book *Leonardo, Psychoanalysis, and Art History*, "Psychobiographical assertions must shoulder the double burden of being not only true but also relevant." I believe the question of whether Leonardo suffered from some form of attention deficit disorder is relevant, but I wouldn't dare to claim that he certainly did. One could argue that on the spectrum ranging from behavioral inhibition to impulsivity, ADHD may be viewed as an extreme manifestation of novelty seeking—a trait that certainly characterized Leonardo.

In his autobiography, Polish American mathematician Mark Kac distinguishes between two types of geniuses:

> In science, as well as in other fields of human endeavor, there are two kinds of geniuses: the "ordinary" and the "magicians." An ordinary genius is a fellow that you and I would be just as good as, if we were only many times better. There is no mystery as to how his mind works. Once we understand what he has done, we feel certain that we, too, could have done it. It is different with the magicians. They are, to use a mathematical jargon, in the orthogonal complement of where we are and the working of their minds is for all intents and purposes incom-

> prehensible. Even after we understand what they have done, the process by which they have done it is completely dark. They seldom, if ever, have students because they cannot be emulated and it must be terribly frustrating for a brilliant young mind to cope with the mysterious ways in which the magician's mind works.

You may think that Kac had Leonardo in mind when he wrote this intriguing passage, but he was describing Richard Feynman, who to Kac was "a magician of the highest caliber."

THREE

And Curiouser

WHEN RICHARD FEYNMAN WAS IN GRADUATE school at Princeton studying physics, a psychology article caught his attention. The author was suggesting that the "time sense" in our brain is somehow determined by a chemical reaction involving iron. Feynman quickly concluded that this was "a lot of baloney"—the chain of reasoning was too fuzzy, and it involved far too many steps, each one of which could have been wrong. Nevertheless, he became sufficiently intrigued by the question itself, what actually *does* control time perception, to start his own series of investigations, even though this problem had nothing to do with the research he was pursuing at the time.

He began by proving to himself that he could count in his head at a standard, roughly constant rate. Then he wondered what affected that rate. At first he thought the rate might have something to do with the pace of heartbeats, but after repeating the experiment while running up and down the stairs (thereby increasing his heart rate), he was convinced the pace had no effect whatsoever. He then tried counting while preparing his laundry list and while reading a newspaper, and neither of those activities seemed to affect the rate. Eventually he realized that there was one thing he definitely couldn't do while counting: he couldn't speak. The reason for this handicap was that he was essentially talking to himself in the act of counting. At the same time, he discovered that one of his colleagues, with whom he had discussed the problem, was counting to himself using a different method, by visualizing in his head a moving tape with numbers on it. This colleague couldn't read while counting but could easily speak. From these seemingly trivial experiments, Feynman concluded that even the simple act of counting to oneself may involve dissimilar processes in the brains of different people: In one case counting primarily meant "talking," while in the other it meant "watching."

Incidentally, in case you're curious, it is now known that there is no single area of the brain that is solely dedicated to recording the passage of time or the body's internal clock. Rather, the system governing the perception of time (and the familiar jet lag) is highly distributed in the brain, involving the cerebral cortex, cerebellum, and basal ganglia. Genes in the liver, pancreas, and elsewhere keep the various parts of the body in sync. People who suffer from Parkinson's disease, for instance, tend to misjudge the passage of time in tasks of time estimation. This topic continues to be an active area of research.

The pattern of wanting to explore every phenomenon that appealed to him continued throughout Feynman's entire life. Alongside his monumental contributions to the quantum theory of electromagnetism and light, to the theory of superfluidity—explaining the peculiar characteristics of the frictionless liquid helium—and to the understanding of the weak nuclear force, which is responsible for some radioactive decays, he relentlessly sought solutions to seemingly mundane, everyday puzzles. His curious mind apparently did not prioritize the problems he chose to tackle. On one day he wrestled with attempting to find a quantum theory of gravity—an extremely demanding problem with which the best physicists are still struggling—and on another, he played with folding strips of paper into origami-like shapes called flexagons. Like Leonardo, he was as intrigued by the formation of waves on the surface of the sea through wind action as he was by friction on polished surfaces. He worked on cutting-edge concepts such as information and entropy (a measure of disorder or randomness) in computer science and on the more prosaic elastic properties of certain crystals. No problem was too small or too dull to address as long as it could be handled in an original way. That's partly why Feynman has been described as the "Sherlock Holmes of physics"; he could solve the most perplexing large or small cosmic mysteries using clues only he could see.

Feynman was not enthralled by science alone. After having a series of arguments about the differences between art and science with his artist friend Jirayr "Jerry" Zorthian, Feynman decided that on alternate Sundays he would give Zorthian lessons in physics, and Zorthian would teach him how to draw. In describing how that agreement came about, Zorthian wrote that Feynman came to him early one morning and said, "Jerry, I have an idea. You don't know a damn thing about physics and I don't know a damn thing about art, but we both admire Leonardo da Vinci. What do you say one Sunday I give you a day of physics, and the next Sunday you give me a day of art, and we both become Leonardo da Vinci." Later, Feynman explained his chief motivation for wanting to learn how to draw: "I wanted to convey an emotion I have about the beauty of the world. . . . It's a feeling of awe—of scientific awe—which I felt could be communicated through drawing to someone who had also had the emotion."

Coming to it from the opposite direction, from an artistic sensibility, this was essentially the same emotion that Leonardo had expressed when he wrote, "Painting compels the mind of the painter to transform itself into the very mind of Nature to become an interpreter between Nature and art."

Art Is I, Science Is We?

After many Sundays during which Feynman was attempting to teach Zorthian physics while learning from the artist how to draw, it became clear that while Feynman was at least making some progress, Zorthian was not. At that point, Feynman wrote, "I gave up the idea of trying to get an artist to appreciate the feeling I had about nature so *he* could portray it. I would now have to double my efforts in learning to draw so I could do it myself." As it turned out, the first drawing by Feynman that was sold at a small art exhibit at Caltech had the rather scientific title *The Magnetic Field of the Sun.* He explained how he had created it: "I understood how the Sun's magnetic field was holding up the flames [of a solar prominence] and had, by that time developed some technique for drawing magnetic field lines (it was similar to a girl's flowing hair)." Isn't this fascinating? Leonardo drew turbulent flows of water like braids of hair, and Feynman drew the magnetic field of the Sun like flowing hair!

Feynman also shared with Leonardo the conviction that knowing the scientific explanation and context of natural phenomena subtracts absolutely nothing from their emotional impact. If anything, he contended, it adds to the impact. He kept coming back to this topic: "Poets say science takes away from the beauty of the stars—mere globs of gas atoms." He was referring to disparaging commentaries such as that made by the nineteenth-century English romantic poet John Keats, who wrote in indignation:

> *Philosophy will clip an Angel's wings*
> *Conquer all mysteries by rule and line,*
> *Empty the haunted air, and gnomed mine—*
> *Unweave a rainbow.*

Naïvely, Keats essentially blamed science for killing curiosity. His contemporary, the mystic poet William Blake, tended to agree: "Art is the tree of life," he wrote, "Science is the tree of death." Blake expressed similar views visually. In a print with pen, ink, and watercolor entitled *Newton* (Figure 9), he portrayed the famous physicist holding a compass. To Blake this compass (which he also used in his depiction of God in the watercolor etching *Ancient of Days*) represented an instrument that constrains the imagination. In the print, Newton himself appears to be so absorbed in his scientific diagrams that he is totally blind to the intricately beautiful rock behind him, which Blake probably employed to symbolize the creative, artistic world.

Feynman could not have disagreed more: "I too can see the stars on a desert night, and feel them," he wrote, "but do I see less or more? The vastness of the heavens stretches my imagination—stuck on this carousel [the revolving Earth] my little eye can catch one-million-year-old light [light that arrives to us from a distance

Figure 9

of a million light-years]. A vast pattern—of which I am a part—perhaps my stuff was belched from some forgotten star. . . . What is the pattern, or the meaning or the *why*? It does not do harm to the mystery to know a little about it. For far more marvelous is the truth than any artists of the past imagined!"

Feynman argued that knowing some of the science behind cosmic objects, phenomena, and events makes us appreciate the beauty of nature even more, and our curiosity about the workings of this magnificent universe is greatly enhanced, not diminished: "Science knowledge only adds to the excitement, the mystery and the awe of a flower." As we shall see in chapter 4, modern psychological and neuroscientific research supports the notion that we become more curious when we know something about a particular topic and feel that there is still a gap in our knowledge to be filled than when we are ignorant about it.

You shouldn't get the impression that Feynman drew only motifs related to physics, such as magnetic fields. In the tradition of most art students, he also sought women models who would agree to pose for him. He described one such experience in his whimsical book *Surely You're Joking, Mr. Feynman!*: "The next girl I met that I wanted to pose for me was a Caltech student. I asked her if she would pose nude. 'Certainly,' she said, and there we were!" As Zorthian has put it, "He was genuinely interested in drawing, but the fringe benefits were the girls, of course."

I know who that student was. She is now a well-known astrophysicist and a good friend, Virginia Trimble, currently at the University of California, Irvine. "Feynman paid me $5.50 an hour for posing, plus he offered to teach me all the physics I could swallow," Trimble told me. They had about two dozen sessions. One of those was scheduled for the day on which Feynman was informed he had won the Nobel Prize in Physics. "He came to tell me that we had to cancel that appointment," Trimble recalled, laughing. Figure 10 shows Trimble at a much later date at her home; one of Feynman's drawings of her hangs on the wall behind her.

I asked Trimble whether during her posing sessions, physics lessons, and conversations with Feynman, she could tell that he was curious about topics that had relatively little to do with the fundamental physics he is best known for. "Of course," she replied. "At one point he became extremely curious about what it was that determined the brightness of a candle. He did not care about any previous attempts to understand this problem—he had to work it out all by himself. Also," she added, "he didn't like silence."

Feynman's sister, astrophysicist Joan Feynman, provided me with an additional insight: "It was easier for him to think about a problem by himself than to read everything that had previously been written about it by others, since the latter required reading also things that were wrong."

Figure 10

The person who might have been Feynman's favorite model, Kathleen McAlpine-Myers, expresses similar sentiments: "I don't know if I could really explain, but he always had this very great curiosity in all situations. It didn't matter what it was—any situation to him was vastly curious, and he just had to know what was going to happen." This attitude of wanting to explore everything by himself is reminiscent of Leonardo's expression: "Though I can not like others cite authors, I shall cite much greater and more worthy: experience." In fact, if not for the elaborate mathematics, a page with Feynman's sketches from 1985 (Figure 11) looks almost as if it had been torn out of one of Leonardo's notebooks.

In spite of the almost five centuries of scientific progress that separate Leonardo and Feynman, the topics that interested them sometimes beautifully overlapped. For instance, both were intrigued by the physics of candlelight. In the manuscript known as

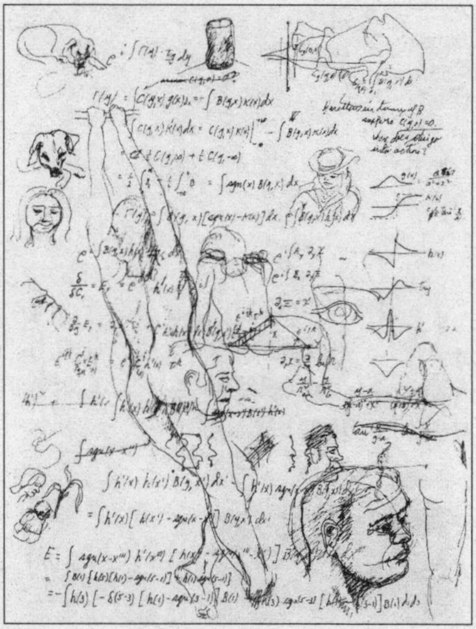

Figure 11

the *Codex Atlanticus* (dating to circa 1508–10), Leonardo devoted a fairly long text to "the motion of the flame." The document summarizes Leonardo's detailed experiments with burning candles and his observations of flickering flames (Figure 12). More important, the text powerfully demonstrates how Leonardo was able to transform his penetrating conclusions from one phenomenon into insights concerning universal principles that he perceived as governing all natural processes. In the words of Leonardo scholar Paolo Galluzzi, Leonardo "recorded the extraordinary chain of thoughts triggered by the candle burning on his table," and his daring analogies carried him to his unified vision of the human body and the physical world.

Figure 12

One "Funny-Looking Picture" Is Worth a Thousand Words

Feynman's most enduring legacy in the world of physics is in the form of cartoon-like diagrams he invented to represent in a pictorial way the interactions among subatomic particles and light. These "funny-looking pictures," as Feynman once described them, are now called "Feynman diagrams." Examples of two such diagrams are shown in Figure 13. Let's first understand what such a diagram really means, at least in very simple terms. The one on the left represents two electrons approaching each other and interacting through the exchange of a "virtual" (unobserved) photon—the carrier of the electromagnetic force. That is, this diagram is supposed to convey the fact that two electrons, both negatively charged, repel each other as they interact through space and time. In the diagram on the right, a neutron and a very light (and very weakly interacting) particle called a neutrino interact by exchanging a virtual W^- particle—one of the carriers of the weak nuclear force—to produce a proton and an electron.

The significance of the visual representation of these fundamental physics processes cannot be overemphasized. Just as Leonardo

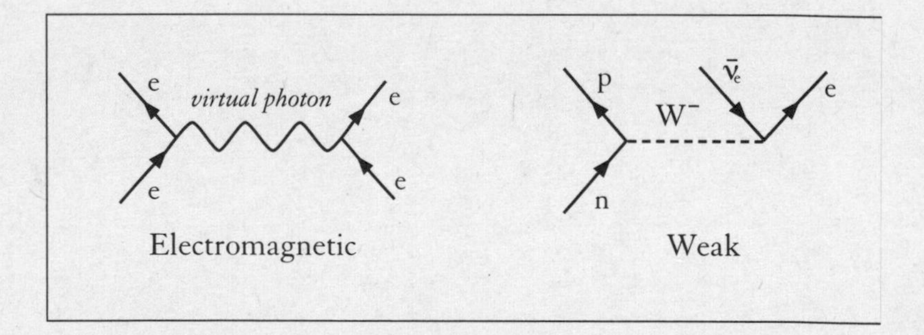

Figure 13

used his unique artistic ability to depict the reality captured by our eyes (at the same time revealing to us something about the workings of his mind), Feynman used his unparalleled physical intuition to produce a new pictorial way to represent the unseen subatomic world. The key point is that the diagrams were not just symbolic cartoons. They provided a precise prescription for how to present and *calculate* the probabilities of all the "virtual" processes that could contribute to the particular interaction being studied and to generate theoretical predictions that could be directly compared with experimental results. For instance, this new method of thinking eventually led to a prediction of the strength of the tiny magnet associated with the electron. That theoretical construct agrees with experimental measurements of the same quantity to within a few parts per trillion!

Feynman diagrams provided physicists with a new, powerful toolkit. From Feynman's perspective, the diagrams also contributed something that was missing from pure calculations: a clear guide on how to proceed in every step—a feature only visualization can generate. In fact, Feynman believed that even Einstein had lost his magic touch once he resorted to calculations alone. He once told physicist Freeman Dyson (and Dyson agreed) that Einstein's great work had sprung from physical intuition, and when Einstein stopped creating it was because he stopped thinking in concrete physical images and became a manipulator of equations.

Richard of All Trades

Even though most of Feynman's seminal works were in physics, he often pondered the relation of physics to other branches of science. He noted, for instance, that theoretical chemistry is in fact the application of the rules of quantum mechanics, and therefore a part of physics, even though making precise predictions in chemistry is sometimes difficult due to the complexity of the systems involved. Turning his attention to biology, Feynman, who was in the Caltech physics department, seriously studied the subject for about a year, with the help of members of the Caltech faculty. He learned enough biology to make an original contribution to the study of mutations in genes. He liked to point out that at their core, life processes, ranging from the circulation of the blood and the transmission of information through the nerves to the way vision and hearing work, are all governed by the laws of physics. These, in essence, were precisely the views of Leonardo, even though Leonardo had no idea what those laws were. In one of his celebrated *Lectures on Physics*, Feynman attempted to explain in some detail the fundamental principles of the workings of enzymes, proteins, and DNA. Recognizing the inherent intricacy of biological components and processes, he nevertheless felt compelled to emphasize that there exists a solid basis for the endeavor to understand life from a physics perspective. In his words, "All things are made of atoms," and therefore "everything that living things do can be understood in terms of the jigglings and wigglings of atoms." As vague as this assertion may sound, to most scientists it represents an indisputable underlying truth.

Feynman was captivated by the fact that astrophysicists had discovered the source of energy that powers the Sun and the stars—nuclear fusion reactions that combine light atoms to form heavier ones, at the extremely hot furnaces in stellar cores. Today astronomy and physics go so intimately hand in hand that the Nobel Prize in physics is occasionally given for discoveries in astronomy.

Astrophysics also provided Feynman with yet another opportunity to voice his opinion that understanding the science underlying natural phenomena amplifies their affecting significance. First, he lamented the fact that poets didn't seem to admire the dazzling knowledge that had been accumulated about planets and stars: "What men are poets who can speak of Jupiter if he were like a man, but if he is an immense spinning sphere of methane and ammonia must be silent?" Second, he went so far as to voice his complaints about poetical work in the pages of the *Los Angeles Times.* In response to his letter, Mrs. Robert Weiner wrote to Feynman that, contrary to his accusations, "modern poets write about practically anything, including interstellar spaces, the red shift, and quasars." She enclosed a copy of W. H. Auden's poem "After Reading a Child's Guide to Modern Physics." Feynman wasn't convinced. In his reply on October 24, 1967, he noted that the poem only confirmed his belief that modern poets "show no emotional appreciation for those aspects of Nature that have been revealed in the last four hundred years."

In this context, Feynman enjoyed telling a story (which may have been apocryphal) that today is sometimes interpreted as having been associated with astrophysicist Arthur Eddington and sometimes with physicist Fritz Houtermans, both of them scientific pioneers who recognized that stars are powered by nuclear fusion "reactors" at their centers. According to this anecdote, Eddington (or Houtermans) and his girlfriend were watching the night sky when she said, "Look how pretty the stars shine!" To which Eddington (or Houtermans) replied, "Yes, and right now I am the only man in the world who knows *why* they shine." The young woman merely laughed at him. The important point here is not whether the story is true or not. Charlotte Riefenstahl, Houtermans's girlfriend and later his wife, was a physicist herself and surely understood very well the importance of deciphering the source of stellar power. The tale is significant because Feynman *thought* it was true, and to him this was just another manifestation of the disturbing lack of recognition and valuing of the "poetry" of science.

Not surprisingly, Feynman pointed to the fields of meteorology and geology as areas in which physicists had *not* been particularly successful in making detailed predictions. In the case of weather forecasts he noted the relatively poor understanding of turbulent flows (a topic in which he showed great interest and which is still largely unsolved), and in the earth sciences he remarked about the gaps in our knowledge concerning both what drives volcanism and the circulating currents within the Earth's interior. In this respect, one of Feynman's characteristics was that he did not hesitate to admit his ignorance: "We do much less well with the Earth than we do with the conditions of matter in the stars." To which he was quick to add with some disappointment mixed with hope, "The mathematics involved seems a little too difficult, so far, but perhaps it will not be too long before someone realizes that it is an important problem, and really works it out." Put differently, he was hoping someone would be as curious as he had always been and would rise to the challenge of attempting to solve that hard problem.

Perhaps the most complex and intriguing topic Feynman touched upon during his tour de force of connecting physics to the other sciences was that of psychology. Here his curiosity manifested itself most dramatically in the following insightful question: "When an animal learns something, it can do something different than it could before, and its brain cell must have changed too, if it is made out of atoms. *In what way is it different?*" Reflecting the sentiments of an era during which there were no techniques such as functional magnetic resonance imaging or experiments with transcranial magnetic stimulation that could deliver images of the working brain, Feynman added, "We do not know where to look, or what to look for, when something is memorized." Even here, however, he half-jokingly but perspicaciously saw a way forward, by first solving a simpler problem: "If we could even figure out how a *dog* works, we would have gone pretty far."

What made Feynman very different from many of his peers was his keen interest not only in many areas of physics but also in subjects far afield. His artist friend Zorthian wrote that he once heard Murray Gell-Mann, a brilliant physicist himself and Feynman's colleague at Caltech, complain about what he regarded as Feynman's many distractions, "We need his [Feynman's] input at Caltech, we need him to talk to us about physics. But what does he do? He goes off and spends all his time with go-go girls and bongo drummers and artists."

You might expect that someone with Feynman's vastly broad knowledge, fervent curiosity, and interest in every domain of fundamental physics would be a strong advocate of what has become known as the "theory of everything"—a framework that would encompass and explain all the elementary subatomic particles and would unify all the fundamental forces of nature. Yet Feynman hesitated. "People think they're very close to the answer, but I don't think so," he admitted. He even wondered about the existence of such a theory: "Whether or not nature has an ultimate, simple, unified, beautiful form is an open question, and I don't want to say either way."

In the end, he recognized that even his insatiable curiosity could reach its limits. Just as Leonardo had to accept that the cave he saw in the mountains might conceal some "marvelous things within it" that were inaccessible, so Feynman acknowledged, "I don't have to know the answer. I don't feel frightened by not knowing things, by being lost in a mysterious universe without any purpose. . . . It doesn't frighten me."

Curiously, there was one other subject in which both Feynman and Leonardo were involved, although given the huge technological gap between their respective times, their association with it manifested itself rather differently. It concerned the simple act of writing.

How Many Angels Can Dance on the Head of a Pin?

Leonardo famously used mirror writing in most of his notes; that is, he started on the right side of the page and wrote to the left, producing text that appeared normal only when reflected in a mirror. We don't know why Leonardo chose to use this particular practice; he did write from left to right in simple notes addressed to others. At least two theories have been advanced, one conspiratorial and the other more practical. The first suggests that Leonardo was attempting to hide his ideas from others, either from people who could steal his inventions or from the Church, whose teachings might have disagreed with his observations. The second theory argues that since Leonardo was left-handed, writing from left to right could have resulted in his hand smudging the wet ink he had just put down.

I should note that Galluzzi is convinced that the conspiracy theory is a red herring. He pointed out that writing from right to left comes very naturally to left-handed people. "In addition," he remarked, "mirror writing is a very silly method to conceal anything, since the text can be easily read with the help of a mirror."

Feynman expressed his interest in the process of writing in a talk he gave in 1959. He opened with a surprising question: "Why cannot we write the entire 24 volumes of the *Encyclopaedia Britannica* on the head of a pin?" Then, with his razor-sharp logic, he analyzed the problem. The estimates were simple enough. Since the head of a pin is one-sixteenth of an inch across, its area is about 25,000 times smaller than the area of all the pages of the *Encyclopaedia Britannica*. Therefore, Feynman deduced, all that would be required is to reduce the size of all the writing in the *Britannica* 25,000 times. Like Leonardo, however, Feynman was not a person to stop after pointing out what the problem was. He immediately proceeded to examine whether that seemed at all feasible based on the laws of physics. He noted that even after such a demagnification, each small dot in a fine halftone reproduction of the encyclopedia would still contain in its area about one thousand atoms, and therefore "there is no question that there is enough room on the head of a pin." He went on to argue that even with the technology of the late 1950s, such a text could be read.

If the *Britannica*, Feynman wondered, why not all the important information that humans have recorded in books throughout their entire cultural history? He estimated the complete documented knowledge could be contained in some 24 million volumes. He therefore concluded that even without any coding, simply reproducing what existed and demagnifying it would take no more than about thirty-five regular pages of the *Britannica*. He conceded that no technique available at the time could do the actual writing, but he insisted that the task was not insurmountable. To further drive that point home, he offered $1,000 to anyone who could reduce the printed page 25,000 times in size, while still keeping it readable.

Feynman was right. The prize was eventually collected in 1985, when Tom Newman, then a graduate student at Stanford University, managed to achieve the desired demagnification using the same technology that is used to imprint electronic circuits on computer chips. He reduced the opening page of *A Tale of Two Cities* to an area of 5.9 x 5.9 micrometers. The resulting text could be read with an electron microscope, reinforcing confidence in Feynman's legendary intuition.

Today's nanotechnology—the manipulation of matter on the atomic or molecular scale—routinely produces fantastic feats of miniaturization. For instance, Joel Yang of the Singapore University of Technology and Design managed to create a tiny copy of Claude Monet's *Impression, Sunrise*, the painting that gave the impressionist movement its name. By swapping oil paints for nanoscale silicon pillars, Yang generated a copy of the masterpiece that is only about one one-hundredth of an inch across. Similarly, the Nano Bible is a gold-plated silicon chip the size of a pinhead on which the entire Hebrew Bible—more than 1.2 million letters—is engraved.

The Last Curiosity

Perhaps the most astounding example of Feynman's incredible inquisitiveness was provided by the moving testimony of his younger sister, astrophysicist Joan Feynman, concerning his final days. In describing that difficult period, she wrote, "So this man who'd been in a coma for a day and a half or something, and hadn't moved, picks up his hands, and goes like this, like a magician, as if to say 'Nothing up my sleeve,' and then he put his hands behind his head. It was to tell us that when you're in a coma you can hear, and you can think."

She added that a little later Feynman came briefly out of the coma and humorously remarked, "This dying is boring, I wouldn't want to do it again." These turned out to be his last words. To Joan, the amazing fact was that to his very last breath, Feynman "was thinking of giving the living some more information about life and nature and what dying was. He was still watching nature, as he was leaving."

Feynman died shortly before midnight on February 15, 1988. Maybe these words of his sum up his personality best: "I don't know anything, but I do know that everything is interesting if you go into it deeply enough."

On October 10, 1517, Leonardo was visited by Cardinal Louis of Aragon. In addition to describing three paintings that Leonardo had shown the cardinal, the cardinal's secretary, Antonio de Beatis, wrote in wonderment about Leonardo, "This gentleman has compiled a particular treatise of anatomy, with the demonstration in draft not only of the members, but also of the muscles, nerves, veins, joints, intestines, and of whatever can be reasoned about the bodies of men and women, in a way that has never yet been done by any other person. All of which we have seen with our eyes. . . . He has, also, written concerning the nature of water, and of divers machines, and other things, which he has set down in an endless number of volumes."

Leonardo died in the Castle of Cloux in France on May 2, 1519. He wrote once, "While I thought that I was learning how to live, I have been learning how to die." Although Vasari's picturesque description of Leonardo dying cradled in the arms of King François I is probably no more than a poetic legend, there is no doubt that the king recognized the full worth of Leonardo's greatness. According to the sculptor and goldsmith Benvenuto Cellini, who was later employed by François I, the king told him he "did not believe that a man had ever been born who knew as much as Leonardo, not only in the sphere of painting, sculpture and architecture, but that he was also a very great philosopher."

Leonardo and Feynman clearly represent the extremely rare, high end on the spectrum of curious people. They both had the ability to turn even human (and indeed their personal) weaknesses into just another interesting piece of the puzzle posed by the grand mystery of the cosmos. However, essentially all individuals (other than perhaps those suffering from very severe depression or brain injury) experience curiosity, even though its depth and breadth may differ from one person to another. In truth, a powerful source of curiosity appears in the world every time a new baby is born.

Are there any immediate, specific lessons we can learn about curiosity in general from this examination of Leonardo and Feynman? At least one seems to be obvious: the brain mechanisms that produce curiosity are apparently neither the ones that are responsible for superior ability in mathematics (which Leonardo didn't have) nor the ones that are in charge of exceptional artistic talents (which Feynman didn't possess). Rather, *a necessary condition for keen curiosity appears to be an information-processing ability.* To be prodigiously curious about as many topics as Leonardo and Feynman were ("he delighted in curiosity about nature," Joan Feynman told me about her brother) requires not only superior cognitive capacity but also brain mechanisms that assign great value to learning and to acquired knowledge. Those, in turn, necessarily entail highly efficient processing of data.

What, then, are the contemporary scientific views on the actual nature, agencies, and goals of curiosity? In chapters 4 and 5 I describe a few of the ideas and experiments that have emerged from advances in modern psychology, and in chapter 6 I give a brief account of fascinating initial results from neuroscience. By their very nature, these three chapters are somewhat more technical than the rest of the book, and they incorporate truly exciting recent findings that have significantly advanced our understanding of curiosity.

FOUR

Curious about Curiosity: Information Gap

UNIVERSITY OF NORTH CAROLINA AT GREENSBORO psychologist Paul Silvia started one of his articles on curiosity and motivation with this sobering observation: "Curiosity is an old concept in the study of human motivation, and like many of psychology's venerable problems, the problem of curiosity seems tractable enough to be intriguing but too complicated to ever solve." What you may find eye-opening is the fact that Silvia's comment was made as recently as 2012. In this light, it should probably come as no surprise that about two decades earlier, University of South Florida psychologists Charles Spielberger and Laura Starr had made a similar remark: "Although many investigators have devoted their efforts to research on curiosity and exploratory behavior, the literature continues to be characterized by diverse theoretical views and contradictory empirical findings." Indeed, the fact that the motivational nature of curiosity has spawned psychological theories that point in different directions suggests that this is a very fluid area of research, in which we still have some way to go before a comprehensive, compelling theory of curiosity will emerge. In fact, curiosity is often bundled together with other psychological elements that characterize human consciousness, and human consciousness, as cognitive scientist and philosopher Daniel Dennett puts it, "is just about the last surviving mystery." What Dennett means is simply that whereas we now know at least how to think about such complex concepts as space, time, and the laws of nature (even though we still don't have a definitive theory for all of them), consciousness "stands alone today as a topic that often leaves even the most sophisticated thinkers tongue-tied and confused."

The problem of fully grasping the nature of curiosity is somewhat compounded by the fact that there isn't even a single definition of the term *curiosity* that is accepted by all. Consequently, phenomena as diverse as the drive to conduct deep-ocean exploration and the emotion evoked by watching *Jeopardy* on TV are often grouped under the same curiosity umbrella. In addition, since neuroscience is a considerably younger discipline than psychology, the precise neural underpinnings of curiosity are even less well understood than the psychological ones.

These difficulties notwithstanding, thanks to recent progress in cognitive psychology and the maturation of neuroimaging techniques, researchers have made and are continuing to make great strides both in their investigations into what stimulates curiosity and the mechanisms it comprises, and in identifying the precise regions in the brain that are activated in the arousal and relief of curiosity.

In order not to get too bogged down with details right from the outset, I shall initially adopt as the definition of curiosity the rather broad formulation suggested by University of Rochester cognitive scientists Celeste Kidd and Benjamin Hayden: curiosity is a drive state for information. Even more simply: curiosity is the desire to know why, how, or who. We shall have to use a sharper, more discerning definition later, especially in relation to cognitive and neuroscientific investigations.

Before probing more deeply into the scientific strands of thought on the essence of curiosity, I want to start with the (at least seemingly) much simpler question: What are people normally curious about in their everyday lives? As a preliminary stab at answering this question, I conducted a small, unscientific poll among a few of my work colleagues. I asked them to describe what piqued their curiosity most, outside of their professional interests. I said I wasn't particularly interested in whether they had once succumbed to the temptation to glance at an open diary; rather I wanted to hear about those themes to which they had actually devoted some time and by which they had been sufficiently captivated to delve into, either through reading and conversations or by web browsing and watching TV programs.

I found the results quite fascinating in that of the sixteen interviewees, no two persons named the same topic. One person was curious about the "nature versus nurture" conundrum, identifying whether heredity or the environment is the main factor that directs and influences people's development and personality. Only two other people mentioned topics somewhat related to this subject. One person was curious about the precise processes in the brain through which children learn; another was curious to know whether there are clearly discernible physiological differences between the brains of "open-minded" people and the brains of those individuals who are extremely rigid in their views. As we shall see, both topics are actually directly linked to curiosity, since one of curiosity's main "goals" is believed to be maximizing learning, and curiosity is also one of several facets of openness. In some sense, therefore, both of these colleagues were curious about curiosity.

Two people were curious about some aspect of sports: one would have liked to know the true extent of doping in various branches of sports, and the other was captivated by the science behind sports. Two people were curious about topics related to the Earth: one about the geological history of our planet, the other about the largely unexplored world at the bottom of the ocean. Two of the subjects involved history: one person's curiosity was focused on World War II, and another's on how we got to where we are today from the time of the Industrial Revolution. The remaining colleagues had their own unique objects of curiosity: antiques, wines, data characterizing people's lives, colors and shapes in interior design, airlines, colony collapse disorder of bees, and the chronicle of achievements of prominent social activists.

Even this unsystematic experiment revealed some interesting points. First, a few of the topics reflected what one might call personal hobbies, that is, interests undertaken primarily for pleasure or relaxation. These included interior design, wines, and antiques. Other subjects seemed to have aroused curiosity because they are surprising or unexpected, for example, the phenomenon of colony collapse disorder—the abrupt disappearance of worker bees from honeybee colonies around the globe—and the stunning revelations about widespread doping in cycling, baseball, and even tennis. Another distinctive feature that appeared to be an originator of curiosity was what MIT cognitive scientist Laura Schulz refers to as "confounded evidence," in other words, situations that are so ambiguous that one cannot decide between different, competing hypotheses or ideas, or where the existing information just isn't sufficient to draw solid conclusions. Themes that fell into this category included the nature versus nurture dilemma and the question of whether open-mindedness and intolerance manifest themselves through something observable in the human brain.

What is most of the U.S. population interested in? For the answer, I examined the most viewed articles on Wikipedia in the years 2012, 2013, 2014, and 2015. Topping the list were tech companies and their social media or information products (such as Facebook, Google, YouTube, Instagram, Wiki), certain blockbuster movies and TV programs (for example, *The Hunger Games, Breaking Bad, The Avengers, The Dark Knight Rises, Star Wars: The Force Awakens*), deaths of famous people (Neil Armstrong, Whitney Houston, Dick Clark, Margaret Thatcher, Nelson Mandela, Robin Williams, Oliver Sacks, Yogi Berra), the lives of celebrities in general (Kate

Middleton, Kim Kardashian, Miley Cyrus), and sports events (such as the 2014 FIFA World Cup).

This unsophisticated internet survey hinted at a few additional elements that can induce curiosity. For instance, the interest in new technology products reflected novelty seeking and a drive to learn. The fascination with the lives (and deaths) of celebrities can perhaps be broadly categorized as "gossip," and gossip (as we shall see in chapter 7) may have played a key role in our evolutionary success. I should note, though, that the Wikipedia list is most likely dominated by the interests of a relatively younger demographic. For example, as of December 2015, 48.5 percent of U.S. internet "addicts" who used Instagram were between the ages of eighteen and thirty-four, and only 5.5 percent were sixty-five or older.

While the extremely diverse nature of the lists of topics that trigger curiosity may look overwhelming at first, psychologists have devised ingenious ways to group such subjects into a smaller number of categories. In particular, recall that psychologist Daniel Berlyne mapped curiosity onto a two-dimensional grid. On one axis curiosity reached from the *specific* (the desire or need for distinct information) to the *diversive* (the ceaseless seeking of stimulation to ward off boredom). The other axis ran from *perceptual* curiosity (aroused by surprising, ambiguous, or novel stimuli) to *epistemic* curiosity (the genuine yearning for new knowledge). Berlyne's insightful classification, while not unique, is useful in the sense that it allows us to locate any particular curiosity on the grid. For instance, one could argue that the curiosity sparked by confounded evidence, or, equivalently, the curiosity that typically motivates basic scientific research, belongs in the epistemic-specific quadrant of the map. That is, we search for some information that will help us decide among alternatives or will guide us in untangling a confusing mess. Ultimately, scientists often conduct their investigations with the goal of finding the answers to particular, clearly defined questions. On the other hand, the curiosity that is driving continuous browsing on Twitter, following up on tabloid headlines, or the desire to check for new text messages more likely maps onto the diversive-perceptual region. In other words, people are searching for some distraction, thrill, or surprise. As we shall see in chapter 6, the distinction between *perceptual* (aroused by novelty) and *epistemic* (the desire for knowledge) curiosity in particular may manifest itself in different brain regions that are activated by curiosity.

Berlyne can be credited with having put the concept of curiosity on the psychological agenda in a big way. A simple comparison of *Psychological Abstracts* prior to 1960, the year in which Berlyne's book *Conflict, Arousal and Curiosity* appeared, with a more recent volume demonstrates his impact on this research field. Later in life, Berlyne, who was also a competent pianist and a huge art lover, also became curious about aesthetics, specifically about the question of what it is that makes certain artworks attractive. In spite of being a rather reserved and shy person, who, as his friends testified, used to stand quietly in a corner nursing a gin and tonic at social functions of psychological organizations, his influence both in the laboratory and on the psychological community at large was unmistakable.

Berlyne made another seminal and enduring contribution to the study of curiosity, with the identification of a class of distinct factors that, in his view, determined whether or not something was interesting and worthy of exploration. These factors are novelty, complexity, uncertainty, and conflict. *Novelty* alludes to topics or phenomena that can't easily be categorized within previous experiences and expectations. For example, the discovery of a new biological species or the first appearance of a smartphone. *Complexity* identifies those objects or events that do not follow regular patterns but rather contain a variety of loosely integrated components. This concept is used, for instance, to describe events in economics, where many people and companies attempt to make sense of market behavior based on whatever information they possess, and where they collectively create outcomes to which they must react rapidly. *Uncertainty* (to which I shall return in more detail in the next section) characterizes situations in which any number of alternative results is possible. Anyone who follows weather forecasts is familiar with uncertainty; despite sophisticated computer models and newfangled technology, meteorologists still occasionally get it wrong. Finally, *conflict* describes circumstances in which new information is incompatible with existing knowledge or biases (as with the discovery that there actually were no weapons of mass destruction in Iraq), or where it isn't clear whether one should respond by taking action or by avoiding the activity altogether. When summarizing Berlyne's work, psychologist Vladimir Konečni appreciatively wrote in his 1978 obituary that Berlyne "wanted to know why organisms display curiosity and explore their environment, why they seek knowledge and information, why they look at paintings or listen to music, what directs their train of thought."

I find it interesting that even the naïve and subjective exercise of polling my work colleagues identified at least two elements that drive curiosity: surprise (which triggers perceptual curiosity) and confounded evidence (which gives rise to a craving for knowledge or epistemic curiosity).

What, then, are the main psychological schools of thought about the causes and mental processes involved in curiosity? (We shall discuss neuroscience in chapter 6.)

Mind the Gap

Like many other trends in modern psychology, a few of the earliest ideas about curiosity were inspired by the work of philosopher and psychologist William James. Presciently, and using present-day cognitive terms, James proposed in the late nineteenth century that what he called "metaphysical wonder" or "scientific curiosity" was a response of the "philosophic brain" to an "inconsistency or a gap in . . . knowledge, just as the musical brain responds to a discord in what it hears." He further suggested that curiosity represents the desire to learn more about those things we don't understand. A century later, Carnegie Mellon psychologist George Loewenstein put forward a contemporary theoretical version of these concepts—an extremely influential framework that has become known as the "information-gap theory."

The basic idea behind this scenario for the explanation of curiosity is simple (once it is pointed out!). It commences with the reasonable assumption that individuals have some preconceived notions about the world surrounding them, or for that matter, about any given topic, and that we seek coherence. When we encounter some facts that appear to be incompatible with our prior actual or imagined knowledge, with our internal predictive model, or with our prejudices, a "gap" is generated. We experience this gap as an aversive state, an unpleasant sensation. Consequently, we are driven to investigate and seek new insights that will reduce the uncertainty and feeling of ignorance. According to this view, curiosity and the ensuing exploratory behavior are not goals in themselves. Rather, they are the means through which we attempt to reduce the uncomfortable sensation caused by uncertainty and confusion. In Loewenstein's own words, curiosity is "a cognitive induced deprivation that arises from the perception of a gap in knowledge and understanding." Simply put, in keeping with the information-gap theory, curiosity resembles scratching a mental or intellectual itch.

Naturally, the information-gap theory identifies *uncertainty*—a perceived disparity between the existing and the desired informational condition—as the chief cause of curiosity. Indeed, being uncertain about possible outcomes at life's challenging junctions can be discomforting. Both in Loewenstein's work and in the context of similar ideas previously expressed by Berlyne, the concept of uncertainty was borrowed from traditional measures of information theory. In simple terms, information theory articulates that all else being equal, situations with a larger number of alternatives or possible results produce a higher uncertainty. For example, if all women's soccer teams are not dramatically different from one another in strength, it will be harder to predict which team will win the World Cup at the beginning of the games than it will be when only two teams are left. Likewise, the uncertainty is larger for potential outcomes of nearly equal probability: if two teams are of comparable skill and motivation, it is more difficult to predict which one will win than if one is decisively superior to the other. Anyone who watched the NBA finals in 2016 between the Cleveland Cavaliers and the Golden State Warriors can testify to the truth of this statement.

A body of research in psychology in the past few decades and in neuroscience in recent years supports at least some aspects of the information-gap theory. For instance, studies have demonstrated that when people are presented with unusual, surprising, or complex objects or situations, those circumstances elicit significantly enhanced attention. Some of those investigations have shown that the desire for inspection and inquiry lasted only until people perceived that they had resolved the uncertainty by acquiring new information. Loewenstein further argues that the magnitude of the gap that people estimate depends on their subjective judgment of the depth of their knowledge and their ability to retrieve information. This is what cognitive scientists refer to as the *feeling-of-knowing*. Loewenstein conjectured that someone with a more intense feeling-of-knowing may deem a certain knowledge gap surmountable where others may not. This perceived capability to overcome a knowledge gap, in turn, was assumed to enhance curiosity, as individuals would feel that with not too much effort they could remove the uncertainty and escape the unpleasant state of anxiety. For example, if someone thinks she knows the names of almost all the actors in a certain movie, she might put extra effort into trying to remember one missing name than if she has no idea who was in the cast.

Loewenstein's information-gap thesis provides a very interesting perspective on the nature of at least some forms of curiosity. In particular, it is easy to see how *specific* curiosity—the urge to acquire a discrete piece of information—can be aroused by an information gap. In any murder mystery, be it a novel by Agatha Christie, Dan Brown, or Robert Galbraith (the pen name of J. K. Rowling) or a film by Alfred Hitchcock, we are curious to know *who* committed the murder, and sometimes also *why* and *how*. Similarly, if your best friend comes to you and says, "I have something extremely important to tell you. Oh, you know what, never mind," this could be truly exasperating. It is straightforward in such cases to identify the information gap that needs to be filled, and curiosity arises because we are fully aware of the precise difference between what we know and what we would like to know. An information gap is also the reason overhearing half a conversation, as when someone next to you is speaking on his or her cell phone, is more curiosity-inducing and more distracting than hearing an entire conversation. In a study by Cornell psychologists, the researchers discovered that listening to such "halfalogues" results in poorer performance on a variety of cognitive tasks that require attention. When we are missing the other half of the story we cannot predict the flow of the conversation, and therefore we find it virtually impossible to tune out the halfalogues. The lead author on the Cornell study, Lauren Emberson, came up with the idea to examine this phenomenon when she was riding a bus to the university for forty-five minutes every day. "I really felt like I couldn't do anything else when someone was on a cell phone," she explained. This may explain, in part, why you see so many people wearing headphones on trains and buses.

The producers of serial TV dramas and soap operas and the authors of thrillers understand the curiosity-arousing powers of information gaps. They attempt to turn the ending of every episode or chapter into a cliffhanger that leaves the audience or readership in suspense.

You'll notice that according to the information-gap scenario, curiosity seeks to satisfy a need that at least superficially appears to be not too dissimilar from physical needs such as those for food, sleep, and the disposal of bodily waste. However, several researchers have pointed out important differences between simple biological urges and curiosity. For example, biological drives such as hunger are typically prompted by clear somatic signals, such as a rumbling belly or stomach pangs. The realization of an information gap, on the other hand, requires a knowledge-based mechanism. To recognize and evaluate the gap, individuals need to actually know something about both their initial information state and their goal, or aspired state. You cannot be too curious about the physical nature of dark energy, for instance, without first knowing something about this mysterious form of energy, which permeates all space and propels the cosmic expansion to speed up.

This raises the first inherent (potential) difficulty with the information-gap scenario, if viewed as a comprehensive theory of curiosity in all of its types and forms. In some cases, it is not easy to see how individuals may be able to properly evaluate their starting or desired levels of uncertainty, given that one never has a complete knowledge of the broader context. It is very common in scientific research, for instance, that the results of one experiment, observation, or theoretical concept raise new questions that had not been anticipated. By way of illustration, Darwin's theory of evolution by means of natural selection brought the question of the actual *origin* of life—a topic Darwin had not touched upon—to center stage. Similarly, the recent discovery that there exist billions of planets orbiting stars other than the Sun has turned the attempts to answer the question "Are we alone in the universe?" into an obsession for many astronomers. The puzzle is, therefore, how does the brain manage to be conscious of and properly define information gaps? How do we assess the extent of our own knowledge and determine how much we do not know? This problem points to a latent clear distinction between biologically determined appetites, which can be felt by everybody in certain circumstances, and curiosity, which may differ from one individual to the next, even under precisely the same conditions. In addition, while specific curiosity may be satisfied if a certain desired piece of information is provided, curiosity in general (in particular epistemic curiosity) and the tendency to explore are actually never truly satiated.

Psychologists have also spotted a few additional problems with the information-gap theory, again if it is taken to be an all-encompassing theory of curiosity. First, it regards curiosity as always being associated with a negative, aversive, unpleasant state. But many experiments on exploratory behavior point to novelty and variety as often being perceived as positive and enjoyable experiences that fuel excitement and attentiveness. In a study with students in seventh and eleventh grades, for example, the students identified as "curious" reported their involvement in school activities as more satisfying and more valuable, rather than being disagreeable. Even uncertainty, the key ingredient driving the information-gap model, does not always have a negative effect, or no one would read murder mysteries or engage in any whimsical activity. While it is undoubtedly true that uncertainty can be discomforting—for example, when one is awaiting the results of a medical test that could confirm or eliminate suspicion of a serious illness—uncertainty about the source of a positive effect could lead to prolonged pleasure.

This last fact was demonstrated in 2005 in an interesting experiment by psychologists Timothy Wilson and Daniel Gilbert and their collaborators. Each participant in the test believed that he or she was one of six students (three women and three men) at different universities, participating in an experiment concerned with impressions formed through the internet. They were told that each one of the six would evaluate the students of the opposite sex, choose one as their best potential friend, and write a paragraph explaining their selection. Every participant was then informed that all three of the opposite-sex students (who were, in fact, fictitious) had selected him or her as best friend. The participants were divided into two groups. Those in the "certainty" group were told which of the three opposite-sex members (supposedly) had written which flattering explanation. In the "uncertainty" group, this information was not provided. Can you guess which group remained happy for a longer period of time? All the participants were pleased to receive the positive feedback about having been selected as a best friend. However, those in the uncertainty group remained significantly more buoyant fifteen minutes later. In other words, if people know that a certain event is positive, they enjoy being curious about it. This is partly why, for example, some expecting parents prefer not to know the gender of the fetus, why the first rush of a love affair is immensely pleasurable, and why some people who record the Wimbledon tennis tournament's final match don't want to know the result before they watch the match. They all enjoy the uncertainty. Only if people are uncertain whether the event will be positive or negative—will they be admitted to a school of their choice? Will a certain medical treatment help?—is the uncertainty regarded as negative.

Fascinatingly, the Romantic poet John Keats introduced the

term "negative capability" to argue that the ability to tolerate and even embrace uncertainty and the willingness to let the unknown remain mysterious are essential qualities for poetic and literary achievement. To Keats, "with a great poet the sense of Beauty overcomes every other consideration, or rather obliterates all consideration." His concept of negative capability has influenced the ideas of a few twentieth-century philosophers, including Roberto Unger, who applied it in social contexts, and John Dewey, who incorporated it into his philosophical tradition of pragmatism. Recall that even Feynman, who was not a poet but a scientist and who generally argued that deciphering phenomena only adds to their beauty, also once said, "I don't feel frightened by not knowing, by being lost in a mysterious universe without any purpose."

A second, related problem that the information-gap theory raises is this: Given that people do occasionally become proactively curious, one could metaphorically invoke a version of Newton's first law of motion, "A body at rest stays at rest," to wonder why anybody would become curious if that is tantamount to asking for an unpleasant sensation. Yet it is very often the case that curiosity about one subject leads to the urge to explore more topics. After all, the most basic characteristic of curiosity is the desire to pose a question, thereby risking generating even more uncertainty, which in the context of the information-gap model is perceived as distressing.

A third problem has to do with the presumed universality of the information-gap theory. That is, even if the basic premise of the theory is correct, at the very least it appears to be an oversimplification, especially when it comes to different types of curiosity. The impression is that there are far too many potential triggers of curiosity for us to be able to collapse them all into just one variable—uncertainty—without some loss of important information in the process. For example, can one truly argue that being curious about the precise nature of gravitational waves, about why music evokes powerful emotions, about how magicians perform their tricks, about what your lunch companion is thinking, about the role of dreams, and about Kim Kardashian's latest Instagrams are all manifestations of a simple information gap?

As we shall soon see, current thinking is that while the information-gap scenario offers an excellent mechanism for certain types of curiosity, curiosity in its most general form comprises a family of mechanisms. Before we discuss other theories, however, there is an additional characteristic of curiosity that requires some explanation, irrespective of what the correct integrative theory (if one exists) may be.

Known Unknowns

In Plato's Socratic dialogue *Meno*, a young and well-born student named Meno attempts to challenge the great Socrates by purporting to demonstrate that inquiring into the unknown is actually impossible. "And how will you inquire, Socrates," Meno asks, "into something when you don't know at all what it is? Which of the things that you don't know will you propose as the object of your inquiry?" Meno points here to the famous "unknown unknowns" problem—things we don't know that we don't know.

The phrase "unknown unknowns" was coined by U.S. Secretary of Defense Donald Rumsfeld during a news briefing held in February 2002 concerning a possible Iraq war. When commenting on the lack of evidence linking Iraq with supplying weapons of mass destruction to terrorist organizations, Rumsfeld told reporters, "Reports that say something hasn't happened are always interesting to me, because as we know, there are known knowns; there are things we know we know. We also know there are known unknowns; that is to say we know there are some things we do not know. But there are also unknown unknowns—the ones we don't know we don't know." In spite of the fact that the comment was perfectly logical, it won Rumsfeld the 2003 Foot in Mouth prize for the most baffling comment by a public figure.

Returning to Meno's apparent puzzle, Socrates chose to reply with an even more bewildering statement, which has become known as "Meno's paradox": "I know, Meno, what you mean; but just see what an eristic [debatable] argument you are introducing—that it is impossible for someone to inquire into what he knows or does not know; he wouldn't inquire into what he knows, since he already knows it and there is no need for such a person to inquire; nor into what he doesn't know, because he doesn't know what he is going to inquire into."

One can paraphrase the last part of Socrates's answer in terms of its application to curiosity, to argue, "She wouldn't be curious about what she knows, since she already knows it; nor about what she doesn't know, because she doesn't know what she should be curious about." Does this mean that we can never be curious? Absolutely not. And this is why Meno's paradox isn't really a paradox.

As far as I know, modern psychologists do not (at least not routinely) refer to Plato's *Meno*. Still, some of them do use a somewhat similar argument to propose that if we examine how our level of curiosity in a particular topic is affected by our existing knowledge in that subject, we would find a function that looks like an inverted U (Figure 14). Simply put, it is very difficult to become curious about something when you know very little about it. Similarly, when you know a lot about a certain topic, you may feel that there is nothing more to be curious about. However, our curiosity is truly piqued when we already have some information about a subject, but we feel that there is still more to be learned. In his provocative answer, Socrates simply omitted that

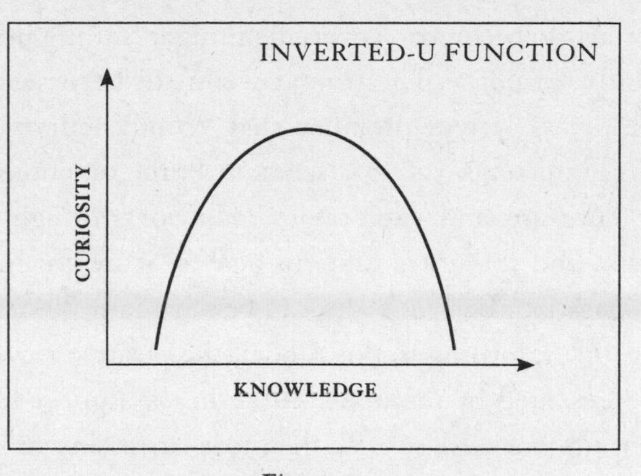

Figure 14

all-important intermediate range of knowledge, what you might call the "known unknowns"—things you know or sense that you don't know.

One version of the inverted-U curve (Figure 15) dates all the way back to Wilhelm Wundt, who in the late nineteenth century was one of the founding figures of psychology. Wundt suggested that as the intensity of a stimulus increases, positive arousal increases too, but only up to a point. For more intense stimuli the ex-

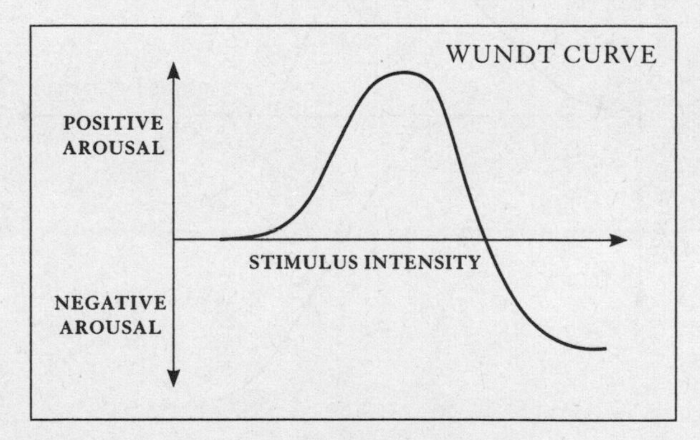

Figure 15

perience starts to become too overwhelming, resulting in a decrease in the positive response. The arousal eventually becomes negative.

In the 1970s, Berlyne proposed that Wundt's curve really represents the interaction of two separate brain functions: one encourages curiosity and exploratory behavior through a reward mechanism, and the other cautions against it by creating an unpleasant sensation. Berlyne's idea can be schematically rendered, as in Figure 16. According to this model, the positive reward mechanism (represented by the upper curve in the figure) acts in such a way that up to a certain level, the more surprising or confounding the phenomenon we observe, the more curious we become. At some point, however, our curiosity saturates, and no matter how much more complex, novel, or puzzling the phenomenon may be, we don't become any more curious—our curiosity levels off (corresponding to the flat part of the upper curve).

In Berlyne's interpretation, the aversive, negative system (represented by the lower curve in the figure) kicks in only at a higher

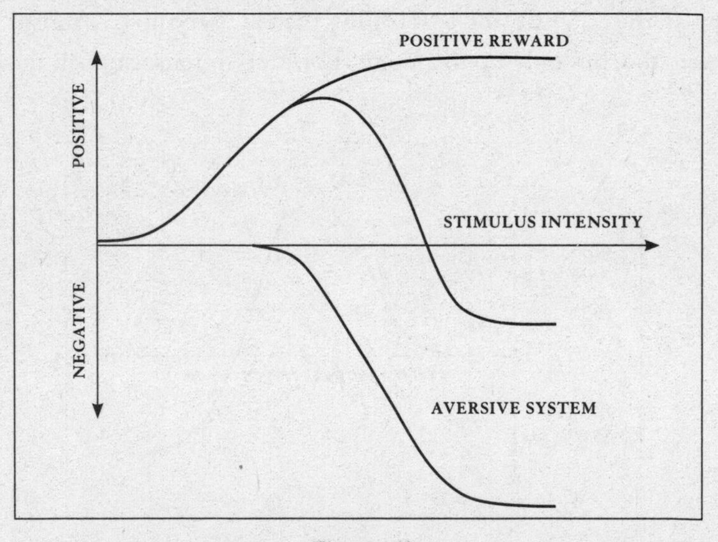

Figure 16

level of prompt intensity, when the stimulus appears threatening or evokes fear. For any stronger stimuli, the negative feelings continuously increase (corresponding to the negative curve dipping lower in the figure). Berlyne suggested that the Wundt curve is simply a consequence of the brain cognitively summing up the positive and negative contributions of the two systems. That is, for as long as the distressing reaction is not activated, curiosity increases as the incentive becomes stronger. Once the brain inwardly starts to weigh the potential negative effects, curiosity starts diminishing, hence producing the inverted-U shape in Figure 15. We can understand Berlyne's concept using a simple metaphor. Imagine that you are traveling through Yellowstone National Park and you suddenly spot a grizzly bear in the distance. This would undoubtedly raise curiosity and excitement. Then, next to the first grizzly you spot a female grizzly bear with her cub. This new discovery generates even more curiosity. Soon thereafter an entire sloth of grizzly bears comes into view in the same area, stimulating an even higher degree of curiosity, especially because bears tend to be solitary animals. But not only doesn't the appearance of more grizzly bears evoke curiosity; fear starts to set in. So many grizzlies in one place is alarming. The concern and fear are further enhanced as more grizzly bears show up nearby.

You may have noticed that Berlyne's version of the cognitive response corresponds almost precisely to the mixture of desire for exploration and fear that Leonardo expressed at the entrance to the cave he discovered in the mountains.

Berlyne's explanation for the inverted-U curve introduced a new element into the theory of curiosity: *a positive reward system.* Interestingly, even though Berlyne's ideas had been presented prior to (and in fact largely inspired) the information-gap theory, that theory still associated curiosity primarily with the need to reduce a negative emotion, with a relatively small (and essentially insignificant) role for positive consciousness. While Loewenstein recognized that exploratory behavior may also be motivated by a positive interest (rather than a feeling of deprivation), his information-gap model implicitly implied that the positive desire for knowledge by itself does not constitute curiosity per se. However, as we shall see in the next chapter, other researchers view curiosity as motivated for its own sake and not just as a vehicle for reducing an unpleasant sensation.

In spite of its thought-provoking attributes, Berlyne's explanation for the Wundt curve has also proved somewhat controversial. First, the interpretation requires the near coexistence of intense antagonistic emotions, pleasure and fear. Opinions vary on whether such a situation is possible, but most psychologists agree that Berlyne's notion that a positive affect should precede the negative one is implausible. For his description to work as conceived, Berlyne had to assume that pleasure is an almost necessary step on the way to a disagreeable state (since aversion kicks in after the positive reward system, at a higher stimulus level, as shown in Figure 16). At least in the case of the emotion of fear, which has been extensively studied by Joseph LeDoux, there is no evidence for the reward system being activated before fear is felt. In addition, at the quantitative level, Berlyne did not provide any compelling explanation for either the relative strengths of the positive and negative emotions or the assumed timing of their activation. Nevertheless, the mere fact that Berlyne considered the possibility that curiosity incorporates both a pleasurable and a discomforting component was significant for the progress in understanding curiosity. This was the seed for the idea that curiosity may consist of a family of mechanisms, which I shall discuss in detail in the next chapter.

As I noted earlier, the information-gap theory raises its own set of difficulties when regarded as a comprehensive theory of curiosity. In addition to the potentially serious problem of identifying curiosity solely with an unpleasant state, the information-gap theory appears, at least at first blush, to fail to explain the common inverted-U pattern. If curiosity is assumed always to intensify with increasing uncertainty, then as we move to higher and higher uncertainty levels, there is no value of uncertainty at which curiosity should start to wane, eventually leading to boredom or even anxiety. In other words, there is no inverted-U curve. This particular aspect, however, has been easily remedied with a relatively straightforward modification to the initial concept: not all subjectively inconsistent observations—not all uncertainties, doubts, or values of the information gap—lead to curiosity. If the gap between what is thought to be known and what is observed is very small, the disparity would not seem (at least in some cases) sufficient for us even to bother with, let alone be curious about. If, on the other hand, the gap is enormous—high levels of doubtfulness or conflict—that may lead to confusion and anxiety, and instead of generating curiosity the gap may be deemed impossible to close. In this interpretation, only an intermediate level of uncertainty can create and sustain curiosity. Put differently, we are not particularly interested in subjects about which we know almost everything or practically nothing. We tend to be interested when we know quite a bit but feel that there is more to be learned (known unknowns). With this simple supplement, the information-gap model accounts for an inverted-U function.

As I shall describe in detail in chapter 6, the expectation (consistent with the information-gap model and the inverted-U shape) that a moderate amount of information increases curiosity but that much additional information reduces it has received support from an interesting neuroscientific experiment.

In spite of its undeniable successes with some aspects of curiosity, the remaining problems with the information-gap model (even when complemented by the inverted-U characteristic) have motivated researchers to turn to different ideas. In an attempt to find other explanations for curiosity, cognitive scientists started to explore the notion that curiosity itself is rewarding and that it is driven by the search for the pleasurable effects of wonder and interest rather than by the unpleasant ones of deprivation and absence of knowledge.

FIVE

Curious about Curiosity: Intrinsic Love of Knowledge

IF CURIOSITY IS NOT A MEANS TO REDUCE THE unpleasantness associated with uncertainty, or at least not only that, what or what else is it? Recent research in psychology suggests that curiosity may provide its own reward. That is, curiosity may be a powerful source of motivation for its own sake, intrinsic motivation, with no control from any external or internal pressure, and with no apparent reward except for the activity itself. The mind, according to this view, should be able to produce rewards that assign value to information gathering and to the acquisition of knowledge.

This perspective has its roots in work done at the beginning of the twentieth century by such psychology pioneers as J. Clark Murray and John Dewey. The concept is based on the simple observation that seeking novel stimuli, interesting people, and new, unexpected ideas appears to be a defining characteristic of being human. Can you even imagine a world without exploration of our external universe and our internal self? Of the microcosm and the macrocosm? Leonardo and Feynman certainly couldn't. In fact, in the same year that Loewenstein published his influential information-gap model, psychologists Charles Spielberger and Laura Starr proposed an optimal stimulation/dual process scenario. In their theory (which, like Loewenstein's framework, incorporated a few of Berlyne's earlier ideas), optimal arousal is achieved as a result of two competing processes. Novel, complex, or incongruent phenomena arouse both a state of curiosity that is perceived as pleasant and an anxiety that is aversive. Spielberger and Starr suggest that when the intensity of the external, triggering stimulus is low, curiosity dominates—along with a desire to explore. At moderate levels of incentive intensity, the fusion of high (pleasant) curiosity and moderate (unpleasant) anxiety tends to trigger specific exploration, that is, a search for distinct pieces of information. Finally, for very powerful stimuli, when we see something that is totally unexpected or extremely confusing, the level of anxiety becomes so high that it motivates total avoidance rather than exploration.

The Spielberger and Starr model reintroduced (following Berlyne) the idea that curiosity can be conceptualized as a positive feeling of interest and wonder. Anyone who ever watched a child with her eyes gleaming as an amateur magician demonstrates his tricks can at least sympathize with this point of view. Adopting a position that was, in some sense, diametrically opposed to that of Loewenstein, Spielberger and Starr identified the unpleasant state caused by uncertainty as "anxiety" rather than "curiosity." Loewenstein, you will recall, suggests that curiosity acts only to alleviate the discomfort associated with an information gap. His model implies that information seeking motivated just by pure interest should not be labeled "curiosity."

Simply put, whereas to Loewenstein curiosity is like scratching an itch to relieve it and the love of learning is something else, to Spielberger and Starr curiosity is the thirst for knowledge, while ambivalence causes anxiety, not curiosity. The important point, however, is that both hypotheses can be put to experimental tests.

As you might expect, Spielberger and Starr's optimal arousal model also leaves a few questions unanswered. The problem is with there being an "optimal" arousal state, which implies that this is a desirable condition to be in. However, if such a state exists, it is not clear why anyone would want to have his or her inquiries, mysteries, and puzzles resolved if that would reduce the positive experience of curiosity to a level of arousal that is less than optimal.

Precisely to avoid these types of problems, and at the same time to incorporate several (sometimes conflicting) ideas into one comprehensive model, psychologist Jordan Litman of the Institute for Human and Machine Cognition proposed in 2005 that curiosity has two aspects. One, which Litman dubbed "I-curiosity," represents the interest (hence the "I") and striving for knowledge that involves pleasurable emotional experiences, while the other, "D-curiosity," results from the feeling of uncertainty and deprivation (hence the "D") associated with not having access to certain information.

I should emphasize that Litman's model is not meant to represent a way to hedge one's bets. He correctly points out that many motivational systems can involve, under different circumstances, both pleasant and unpleasant emotions. For instance, hunger may be stimulated by watching a TV commercial for Doritos or films such as *Babette's Feast, Mostly Martha*, or *Julie & Julia*, which are all celebrations of exquisite cuisine. Alternatively, you may realize that you're hungry from the pangs caused by an empty stomach or the desire to pamper yourself when you feel neglected. Similarly, the desire to engage in sex may be sparked by a spontaneous, pleasant emotion toward a loved partner or by the deprivation induced by a long absence, such as military service in a foreign country.

To put it differently, according to Litman's conjecture, curiosity may be both a reduction of an aversive state and an induction of an intrinsically motivated enjoyable state. Which one dominates will depend on the type of stimulus and perhaps on individual differences. For instance, the beating of the human heart, which triggered a torrent of epistemic curiosity (the drive to explore) in Leonardo and caused him to fill innumerable pages with notes, hardly even registered with many of his contemporaries. Similarly, not remembering the names of the students who sat next to them in high school may drive some people crazy and leave others totally indifferent. Or seeing an unfamiliar animal in a zoo may evoke perceptual curiosity in some visitors (they will look for the placard identifying the animal) and epistemic curiosity in a few others (they will extensively read about it at home).

This general idea of curiosity comprising a family of mechanisms rather than representing a single process has been further examined by a team of researchers led by Jacqueline Gottlieb of Columbia University, Celeste Kidd of the University of Rochester, and Pierre-Yves Oudeyer of the French Institute for Research in Computer Science and Automation. They suggest that the weight we assign to the different components and forms of curiosity depends on both the stimulating event or topic and the individuals themselves (in terms of their knowledge base, biases, and cognitive characteristics). As we shall see in chapter 6, recent results from neuroscience support a scenario in which different kinds of curiosity involve distinct brain regions.

As I have pointed out, individual differences in curiosity can be enormous. Whereas Leonardo and Feynman, for instance, were curious about almost everything, some people have very few interests outside their work. These differences have traditionally been studied largely in the context of a general trait labeled "openness to experience," considered one of the "Big Five" dimensions of human personality. In psychology, those Big Five personality attributes are openness to experience, conscientiousness, extroversion, agreeableness, and neuroticism (forming the acronym OCEAN). Of these five characteristics, openness to experience is the one that is assumed to encompass intellectual curiosity and preference for novelty and exploration, even though the precise definition of *openness* is somewhat controversial. Broadly speaking, people with high openness are not only more curious but also more appreciative of such things as complex forms of art. They have a higher capacity to think in abstract terms.

Even if we accept the very reasonable idea that curiosity (in all of its different manifestations) involves both a deprivation induced by uncertainty and an anticipation of reward stimulated by an intrinsic striving for knowledge, many things remain unknown. How exactly does the brain put value on knowledge and on its acquisition? What is the mental strategy (if there is one) that underlies information seeking and exploration? We know, for instance, that the white noise on a TV screen when there is no transmission contains a vast amount of information. Yet I am not aware of anybody who is riveted by those flickering points of light and accompanying hissing noises. What is the process in which the human mind sifts through all the information that bombards us and decides what to be curious about?

Cognitive scientists are trying to understand whether curiosity-induced behavior has any strategic plan or ultimate goals.

Explore All Options

Both everyday experience and numerous studies demonstrate that individuals engage in exploratory behavior—part of what we normally think of as curiosity—even in the absence of any financial or other obvious external rewards. Common wisdom has it that the activities people tend to concentrate on follow a pattern: they avoid challenges that are either too easy and therefore perceived as boring, or too difficult and therefore seem intimidating and frustrating. How, then, do people channel their curiosity and how do they organize their exploration if they are free to choose among a large number of paths and options? As we know, many activities may lead to cognitive dead ends or to incomprehensible situations. For instance, a young boy should not choose James Joyce's *Ulysses* to read as his very first book, and a young girl curious about how the brain functions should not start by performing brain surgery.

Neuroscientist Jacqueline Gottlieb and her collaborators carried out a few fascinating experiments to examine the question of whether our brains use some universal strategy to guide curiosity in intrinsically motivated open-ended exploration. The researchers asked fifty-two subjects (twenty-nine women and twenty-three men) to choose which short computer games they wished to play. There were two sets to choose from, and the difficulty of the games varied within each set.

The results were quite surprising. Gottlieb and her colleagues found that in spite of the fact that there was no external guidance and no tangible reward, the subjects spontaneously organized their exploration in a consistent pattern. First, the participants were sensitive to the difficulty of the tasks: they diligently started with the easiest games and advanced to the more difficult ones. Second, the subjects were interested in exploring all the available choices: they sampled the entire set of games, including those that were so difficult that they were essentially impossible to master. Third, the subjects tended to repeat the games of moderate to high difficulty. Finally, the participants liked novelty, which they injected into the experience by selecting new games, but they also preferred to choose games at a level of difficulty they had already become familiar with.

These findings have interesting implications for the nature of epistemic curiosity (the craving for knowledge). First, the fact that the participants explored even the most demanding tasks and experimented with novel sequences suggests that people do seek to acquaint themselves with the entire accessible landscape of alternatives. They attempt to increase and mentally encode their knowledge and to enhance their ability to make reliable predictions about new opportunities. This characteristic has been dubbed "knowledge-based intrinsic motivation," and its important function is that it helps to reduce prediction errors. A high school student reviewing information about many universities before deciding where to go is being driven by knowledge-based intrinsic motivation. At the same time, the two other findings—that participants repeated challenging games and selected novel sequences only for games in which they had performed well—suggest an inherent desire to excel through practice. This is called "competence-based intrinsic motivation."

Gottlieb's results provide us with a few important insights into the way epistemic curiosity operates under open-ended circumstances. Perhaps the most surprising finding is that even with no hints, pointers, or guidelines, people tend to follow a similar path. In terms of its strategic plan, epistemic curiosity appears to aim at two goals: to act as a motivator for us to understand the limits of potential choices and, more important, to maximize knowledge and competence.

Since Gottlieb is one of a surprisingly small number of researchers whose main research focus is curiosity, I naturally became curious to know what attracted her to this topic. "I started from trying to understand the mechanisms of attention," she told me, "and then I was pushed towards curiosity from two different directions. First, from behavioral considerations, I was interested in what role attention serves in directing our behavior."

"What exactly do you mean?" I inquired.

"Most studies that follow eye movements as indicators of attention, for instance, ask the subjects to pay attention to something such as a red square on a screen, and then the researchers examine how this directed attention modifies things such as reaction times. However, they usually don't study how the actual decisions are made, that is, what makes something worthy of attention." After a short pause she continued: "So I decided that we have to investigate the logic that guides that type of selection. For example, we often select in relation to an expected reward. This is known as goal-directed behavior. But there are still many things that we are interested in that promise no obvious rewards. This is where curiosity comes in." She then added, "I wanted to know which process is involved in curiosity, what directs us to *learn* even when we don't know what the precise consequences of this learning are going to be."

"And what was the second thing that directed you to curiosity?"

Gottlieb laughed. "You didn't forget that there was a second thing. That came from neuroscience. I wanted to know which areas in the cerebral cortex [the outer layer of neural tissue in the brain that is central for consciousness] select the stimuli that are being attended to. There are many models of brain responses, and they again usually explain situations in which the subjects have a goal or a reward in mind. Just as in the behavioral case, I was more interested in those 'goal-independent' selections. So I had a convergence onto curiosity from both the behavioral aspect and neuroscience."

I was still curious about Gottlieb's own path to scientific research, so I asked, "Was there anything in your background that you think influenced your decision to become a scientist?"

"Ultimately, I think that this is the occupation that best captures my skills. In high school I wanted to be a pianist, but then I realized that my talents as a pianist were somewhere in the middle of the distribution and that it would have been very hard for me to make a mark. Then, while studying at MIT, I discovered that I have a natural ability to carry out analytical work. I love the creativity and freedom that come with science. I have a very low tolerance for boredom, and science is the one discipline in which there are always new challenges." After a brief silence she added, "My greatest joy is when I learn something new."

This is precisely what defines an intellectually curious person.

Gottlieb's experiments were performed with adults. There used to be a joke in research circles that all the subjects in psychology experiments are always college freshmen or sophomores, and therefore all the results and findings in psychology apply only to that demographic. In recent years, however, much more attention has been given to those tiny "curiosity machines"—children, toddlers, and even babies—in a quest to understand whether the curiosity manifested by infants and children is similar to that observed in adults. Do perceptual, epistemic, diversive, and specific curiosity remain stable across the life span, or do they change with age? Though there aren't yet enough longitudinal studies that directly compare children and adults, research in the past two decades is painting a more coherent picture of curiosity in children. What follows are a few examples of what I regard as compelling experiments in this fascinating area.

Out of the Mouths of Babes

If you've ever watched a ten-month-old playing with one of those Shake & Rattle bead balls, you know that she will jiggle the toy side to side, put it in her mouth, bang it on the floor, and try to separately slide some of the colorful pieces. This may last for a few minutes, until she glimpses a board book nearby. Turning her attention to that, she will put it in her mouth and then clumsily attempt to turn the thick pages one at a time. What is it that ignites the child's curiosity?

Laura Schulz is a cognitive scientist at MIT's Early Childhood Cognition Lab. She and her collaborators have spent the past decade or so attempting "to figure out how children learn so much from so little so quickly." Indeed, within just a few months, children learn a rich complement of motor skills, recognize their parents, and start to interact and communicate in a variety of ways. The attentional mechanisms of infants must somehow select from their immediate rather complex environments those elements that make their learning process both efficient and tractable. Schulz and other cognitive scientists are trying to understand how children manage to "draw rich inferences from sparse, noisy data."

There is considerable evidence, much of which has come from pioneering experiments by Harvard psychologist Elizabeth Spelke, that infants start their lives with a few simple heuristics, or ways of solving problems on their own, that guide their initial exploration. Spelke studies babies because "adult minds are already too full of facts," she told me in a phone conversation. "It's best to determine what we know at birth." To penetrate into the minds of infants, she recognized that the length of time babies spend staring at things is an excellent indicator of what triggers their curiosity. For instance, the onset of motion attracts their gaze, as do areas characterized by high contrast, and also human faces. All of these are rich in information value. Detecting motion is an obvious evolutionary necessity for survival, and contrasts help in distinguishing discrete objects and in recognizing their shapes. Additionally, babies know that if they grab the leg of a doll, the rest of the doll will come with it—all parts of a distinct object move together. They also know that solid objects cannot pass through other solid objects, and they have an innate sense of number and the geometry of the space surrounding them. The bias toward human faces is part and parcel of developing social skills, affectionate personal relationships, and eventually the language capacity. Spelke and colleagues Katherine Kinzler and Kristin Shutts have also found that infants show a clear social preference for people who speak in a language and accent the infants have already become familiar with. This was found to be true for both American and South African children, even though the latter are living in a more diverse linguistic environment.

Conditioning tests, in which infants have to anticipate and react to repeated events, show that they also seek information that will help them to generate predictive strategies. But can we call these first signs of attention biases "curiosity"? That depends on the precise definition of the term. Under the very broad definition that I adopted earlier to initiate the discussion ("a drive state for information"), these infant heuristics certainly qualify as expressions of curiosity, as would reactions to games of peekaboo or Pop! Goes the Weasel. One may correctly argue, however, that this definition encompasses everything that happens from the moment we first open our eyes, not only the state of being genuinely curious. If, therefore, for something to be called curiosity we insist on its being represented by a clear understanding of the initial and the desired information states, then these early, low-level attentional behaviors cannot be considered curiosity. Perhaps they are its precursors. Be that as it may, if we are interested in inquisitiveness beyond the very basic innate heuristics, how do children select the subjects to which to direct their curiosity during the period in which their mental perception of the world is evolving?

In experiments conducted at the University of Rochester with seven- and eight-month-olds, Celeste Kidd and her collaborators measured the infants' visual attention to sequences of events of varying complexity shown on a screen. The researchers discovered that the probability for the infants to look away from the screen, indicating a loss of interest, was highest for sequences whose complexity was either very low or very high. In other words, the researchers identified a "Goldilocks effect": the infants directed their curiosity to sequences that were neither too simple nor too complex (an inverted-U-type preference). Recall that the same effect was found in Gottlieb's experiments with students playing computer games.

Kidd's results seem to imply that the brain of an infant adopts a strategy that allows it not to waste valuable cognitive resources on either irreducibly complex or easily predictable phenomena. This interpretation suggests that even in infants, curiosity depends on the initial state of knowledge and on one's expectations, and that it acts to maximally enhance one's learning and encoding potential.

A different set of experiments at MIT reveals another interesting aspect of curiosity in children. It shows that, like adults, children organize their play and exploration with the goal of reducing uncertainty and uncovering the true causes of phenomena. This was demonstrated in a simple jack-in-the-box-type experiment, designed by cognitive scientists Laura Schulz and Elizabeth Bonawitz. The researchers presented preschoolers with a red box with two levers. When a researcher and a child simultaneously pressed down on one lever each, two small puppets popped up at the very center of the top of the box. There was no way to tell which lever caused which puppet to pop up, or even whether only one of the two levers raised both puppets. Hence the evidence was *confounded.* The researchers repeated the experiment with a second group of children, only this time the conditions were deliberately unconfounded—the child and the researcher either took turns pressing the levers, or the researcher demonstrated to the child how each lever acted separately. In this instance, therefore, the child could precisely tell which lever operated which puppet. Following the demonstration to both groups, the researchers introduced a new, yellow box and left the children alone to play. The results were truly fascinating. The children in the "confounded evidence" group tended to continue to explore the red box until they had figured out its operation. The children in the "unconfounded" group showed the expected preference for novelty and immediately directed their attention to the new yellow box.

What these and the results of other experiments seem at least to suggest is that curiosity in children is often related to maximizing learning and to the discovery of the causal relationships that govern the child's environment. Put differently, children search for a way to explain everything in a step-by-step account. If this inference is correct, however, then it also makes for a very clear and interesting prediction: children's curiosity should especially be piqued by and focused on exploring those situations in which their expectations are violated. This prediction can be tested by examining how exploration and learning are affected when the observed evidence contradicts prior beliefs.

Bonawitz, Schulz, and their colleagues attempted to do precisely that through a series of extensive studies. In one carefully planned experiment, the researchers asked children to scrutinize nine *asymmetric* blocks of Styrofoam that could be stabilized on a balancing rod. In an initial "belief-classification" task, the researchers closely observed whether the children were attempting to balance the blocks at their *geometric* center, in the middle of the block, or at their perceived *center of mass*, closer to the heavier end (Figure 17). The experimenters took hold of the block just before the children could stably set it on the post so that the children did not get a chance to actually see whether or not the block was balanced. In this way, the researchers created a group of children (with a mean age of six years and ten months) with a known prior bias toward the geometric center as the balancing point, and a group of somewhat older, more experienced children (mean age of seven years and five months) with a prior belief in the center of mass as a balancing spot. They also had a group of younger children (mean age of five years and two months) who had no prior "theory" about the balancing point and who therefore tended to balance blocks simply by trial and error.

In the second stage, all the groups were shown blocks that appeared to be in perfect equilibrium on the rod. That, however, is when things started to get interesting. Children with "geometric center" and "center of mass" theories who were shown identical,

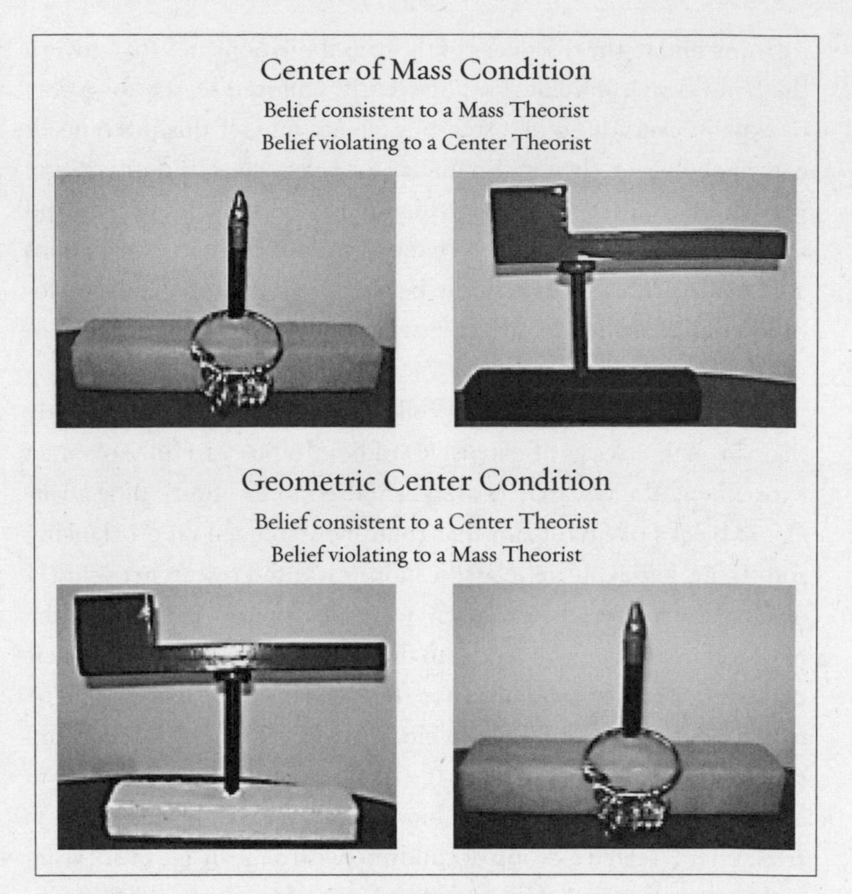

Figure 17

balanced configurations explored the blocks differently, depending on their prior beliefs. When the children were shown a block balanced at its center of mass (consistent to the center-of-mass "theorists" but belief-violating to the geometric-center "theorists"), those who had their belief defied spent more time exploring the block, while the others preferred to examine a new toy. The behavior of the two groups that had prior theories was reversed when the block was balanced at its geometric center. The children who had no prior theory always preferred the new, untried item, irrespective of the evidence presented to them.

In related experiments, the researchers showed the children that the precisely balanced blocks were actually held in their place by a magnet. The reactions of the different groups were again interesting. Both the geometric-center group and the center-of-mass group used the new element—the magnet—in an attempt to explain the evidence, but only in those cases in which their prior beliefs had been at odds with the new observations. That is, geometric-center theorists who saw the block balanced at its center of mass concluded that this was only because the block was held in place by the magnet. The same was true for center-of-mass believers who were presented with a block balanced at its geometric center. Moreover, in experiments in which the presence of the magnet was not revealed, the children used the new evidence of the belief-violating balanced block as a motivating force to rethink and revise their predictions. They did not feel compelled to change their beliefs if an auxiliary explanation (in this case, the presence of the magnet) was available.

Overall, the picture that emerges from all the studies with children is one in which the components of curiosity that are directed to newness or unfamiliarity and to the engagement with purely pleasurable stimuli (in other words, diversive and perceptual curiosity) sometimes take second place to the desire to maximize learning, to understand cause and effect, to discover the structure of the world, and to decrease prediction error (that is, epistemic curiosity).

Studies show that before babies are nine months old, while they are very good at handling and mouthing objects, can distinguish between familiar and strange, and are very alert to sights and sounds, they rarely show interest in somebody else's wishes or intentions. However, within a very short period of time, infants develop a new kind of mental relationship with the outside world. It becomes of primary interest to them.

Experiments with 1,356 men and 1,080 women ages seventeen to ninety-two have shown that novelty seeking (and more generally, perhaps, some aspects of diversive and perceptual curiosity) tends to decline with age, while specific and epistemic curiosity appear to remain stable into adulthood, and even into old age. Put differently, being "infovores" and *wanting* to learn is a constant characteristic of humans, but the willingness to take risks for novelty, thrill, or adventure and the ability to be surprised decrease as we grow older.

Cognitive scientists and psychologists attempt to decipher the intricacies of the operation of the human mind when we are curious. However, our understanding of curiosity cannot be complete without a complementary comprehension of the associated physiological processes inside the human brain.

SIX

Curious about Curiosity: Neuroscience

SINCE THE EARLY 1990s, NEUROSCIENTISTS HAVE added a powerful new tool to their research arsenal, one that literally enables them to *image* curiosity in action in the brain. Functional magnetic resonance imaging (fMRI) is a procedure that allows researchers to examine which regions of the brain are activated during particular mental processes. The technique relies on the fact that when a certain area of the brain is used intensively, the energy required for the neural activity results in an increase in the blood flow into that region. The working brain can therefore be mapped in detail by taking snapshots of the changes in the blood flow, using the blood-oxygen-level-dependent (BOLD) contrast—the fact that oxygenated blood has different magnetic properties from deoxygenated blood and that the relative difference can be imaged. When combined with supplementary cognitive research, fMRI offers a new dimension to the studies of curiosity. A few neuroscientific experiments have been particularly innovative and influential in advancing our understanding of the neurophysiological underpinnings of curiosity.

Jeopardy in the Brain

In a seminal investigation in 2009, Caltech researchers Min Jeong Kang, Colin Camerer, and their colleagues used fMRI with the goal of identifying the neural pathways of curiosity. The scientists performed a test in which they scanned the brains of nineteen people with fMRI while these folks were presented with forty trivia questions. The questions, on various topics, were specially selected so as to create a diverse mixture of high and low specific-epistemic curiosity, that is, interest in specific knowledge. One question asked, "What instrument was invented to sound like a human singing?" Another, "What is the name of the galaxy that Earth is a part of?" The participants were asked to sequentially read a question, guess the answer (if they didn't actually know it), rate their curiosity to find out the correct answer, and indicate how confident they were in their guess. In the second stage, each subject saw the question presented again, immediately followed by the correct answer. (In case you are curious, the answer to the first sample question is the violin; to the second, the Milky Way.) The reported curiosity was found to be an inverted-U-shaped function of the uncertainty.

The fMRI images showed that in response to self-reported high curiosity, the brain regions that were significantly activated included the left caudate and the lateral prefrontal cortex (PFC)—areas that are known to be energized on anticipation of rewarding stimuli (Figure 18). This anticipation is the type of feeling you have before the curtain goes up on a play you have wanted to see for a long time. The left caudate had also been shown to be activated during acts of charitable donation and in reaction to punishment of unfair behavior, both of which are perceived as rewarding. Kang and her colleagues' findings were therefore consistent with the idea that epistemic curiosity—that is, the hunger for knowledge—elicits anticipation of a reward state, which indicates that the acquisition of knowledge and information has value in our minds. Somewhat surprisingly, though, the brain structure known as the nucleus accumbens, which is thought to play a central role in the reward and pleasure circuits (and is one of the most reliably activated regions in

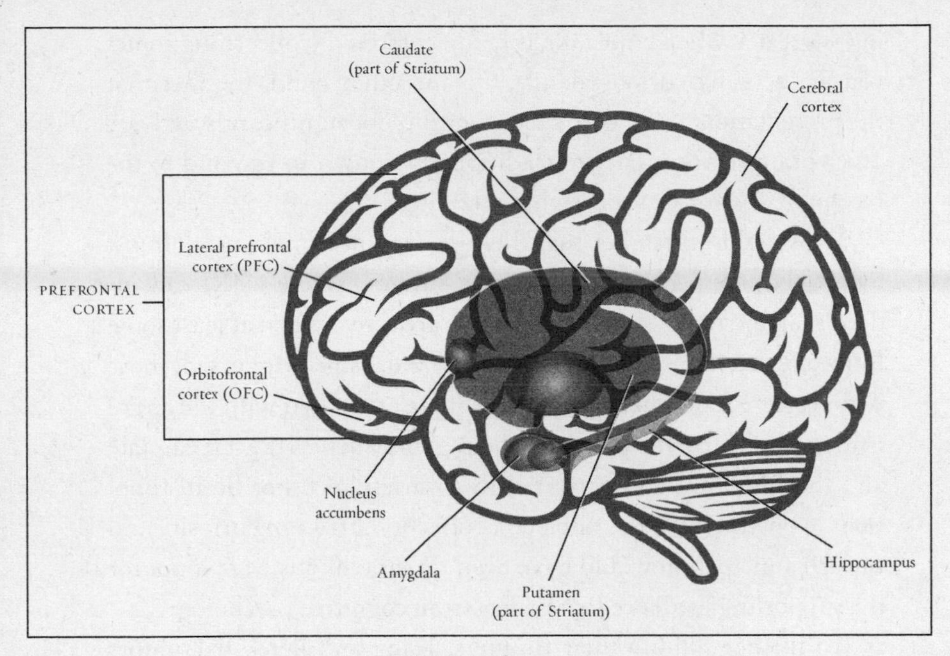

Figure 18

reward anticipation), was not activated in Kang and her colleagues' experiment. The researchers also found that when the correct answer was revealed to the subjects, the regions of the brain that were significantly energized were those typically associated with learning, memory, and language comprehension and production (such as the inferior frontal gyrus). Notably, activations were found to be more powerful when the participants were shown the answers to questions for which they had previously guessed incorrectly than following correct guesses. The subjects also exhibited enhanced memory of correct answers when they had initially been wrong. A subsequent behavioral study showed that a higher curiosity in the first session correlated with better recall of surprising answers even ten days later. This result could perhaps be expected, since the information is considered more valuable and the potential for learning is greater when a mistake is being corrected (concerning topics you are actually curious about). On the other hand, the fact that the presentation of the correct answer did not significantly activate other brain regions that are traditionally known to respond to the receipt of reward was somewhat puzzling.

We should remember that there is one uncertainty that almost inevitably plagues all neuroimaging studies. While fMRI can indeed map the regions of the brain that are active when at least some form of epistemic curiosity is induced (and, as just discussed, those regions were found to be the ones that are associated with the anticipation of reward), those very same regions (such as the left caudate and the PFC) are also activated in a variety of other brain functions. Consequently, the inferred connection between curiosity and reward anticipation would have been rather tenuous were it not for the supporting evidence that comes from cognitive psychology.

To further solidify their findings, Kang and her collaborators performed an additional test, crafted so as to enable a distinction between true reward anticipation and the simple function of increased attention (which in previous experiments had also been found to activate the left caudate). The new experiment had two components. In one, the researchers allowed the subjects to spend at any time one of twenty-five tokens to find the correct answer to one of fifty questions (ten questions were added to the original forty). Since the number of tokens equaled only half of the number of questions, the implication was that by spending a token on a particular answer, subjects were opting to give up on another. In a second condition of the experiment, subjects could decide to wait between five and twenty-five seconds for the answer to appear, or they could quit waiting and skip to the next question, thereby missing out on the correct answer to the preceding question. Both actions (spending a token or waiting for an answer) came at a certain cost, either of resources or of time. The results showed that spending tokens or time was strongly correlated with the expressed curiosity. This outcome considerably strengthened the interpretation of curiosity as an anticipation for reward since people are generally more inclined to invest (either time or money) in items or actions they expect to be rewarding.

Overall, in spite of the remaining uncertainties, the pioneering work of Kang and her colleagues did suggest that specific-epistemic curiosity is linked to anticipation of information that is viewed as a reward. The additional findings, which demonstrated a strengthening of memory in response to being initially curious but wrong, indicated that curiosity enhances the potential for learning. As I shall discuss in more detail later, this finding may provide important clues for improving teaching methods and for communicating information more effectively.

As groundbreaking as Kang and her collaborators' work was, however, it left many questions unanswered. In particular, this study explored only one kind of curiosity—specific-epistemic—the one that is expected to be evoked by such knowledge-based catalysts as trivia questions. Does the brain respond similarly to the stimulus of novelty, surprise, or the simple desire to avoid boredom? Does the response depend on the form of the stimulus? For example, are the processes in the brain the same when we become curious by examining an image rather than by reading a text? A study published in 2012 attempted to address a few of these intriguing questions.

Blurry Images

Scanning people's brains while they are being curious certainly provides for an exciting experiment. But how exactly do you ask someone to be curious? Even requesting that the participants rate their curiosity (say, on a 1–5 scale) is sure to introduce a certain amount of subjective ambiguity. Cognitive scientist Marieke Jepma of Leiden University in the Netherlands and her team used a different method from that of Kang and colleagues to pique the curiosity of her subjects. Specifically, Jepma decided to focus her attention on *perceptual* curiosity—the mechanism aroused by novel, surprising, or ambiguous objects or phenomena. The idea was to fan the embers of curiosity with equivocal stimuli, the type that are open to many interpretations. The researchers therefore scanned (using fMRI) the brains of nineteen participants who were shown blurred pictures of various common objects, such as a bus or an accordion, which were difficult to identify because of the blurring. To manipulate the triggering and relief of perceptual curiosity, Jepma and her colleagues cleverly used four different combinations of blurry and clear pictures (Figure 19 illustrates the set of combinations): a blurred picture followed by its corresponding clear picture; a blurred picture followed by a totally unrelated clear picture; a clear picture followed by its corresponding blurred picture; and a clear picture followed by an identical clear picture. The subjects, therefore, never knew what to expect or whether their curiosity about the identity of the object would be relieved.

Since Jepma's study was one of the very first experiments that attempted to demonstrate the neural correlates of perceptual curiosity, the results were sure to generate great interest, and they did not disappoint. First, Jepma and her collaborators discovered that perceptual curiosity activated brain regions that are known to be sensitive to unpleasant conditions (even though not exclusively to those). This was consistent with expectations from the information-gap theory—perceptual curiosity appeared to produce a negative feeling of need and deprivation, something akin to thirst.

Second, the researchers observed that the relief of perceptual curiosity activated known reward circuits. These findings were again consistent with the idea that the termination of the distressed state

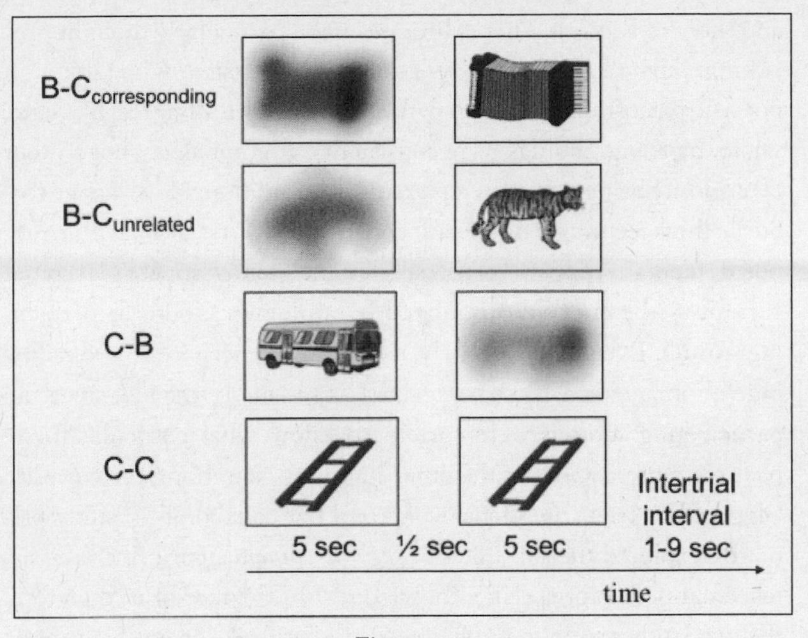

Figure 19

that typifies perceptual curiosity—by providing the desired information—or at least the reduction in its intensity, is perceived by the mind as rewarding. Simplistically put, being perceptually curious is a bit like being deprived, conflicted, or hungry. Satisfying one's curiosity is comparable to having good food, good wine, or good sex.

Jepma and her collaborators uncovered a third interesting fact: the induction and reduction of perceptual curiosity acted to enhance incidental memory (memories formed without really trying), and they were accompanied by the activation of the hippocampus (Figure 18), a brain structure recognized to be associated with learning. This discovery provided additional support to the conjecture that igniting curiosity is a potent strategy not only to motivate exploration but also to strengthen learning.

The differences, rather than similarities, between Jepma's results and those of Kang and her colleagues were particularly thought provoking. Jepma's discoveries were generally consistent with (although not a proof of) curiosity being fundamentally a disagreeable state, while the Kang findings were consistent with (but again, not a proof of) curiosity being primarily a pleasurable condition. How can we reconcile these seemingly discrepant conclusions? First, as I have already noted, Jepma's study was expressly designed to investigate perceptual curiosity—the curiosity stimulated by ambiguous, odd, or perplexing stimuli. Even more precisely, the curiosity mechanism evoked by blurred images can be characterized as *specific-perceptual*, since the participating subjects were curious to know what particular fuzzy pictures represented. On the other hand, by examining the curiosity triggered by trivia questions, Kang and her collaborators' study primarily explored the substrates of *specific-epistemic* curiosity—the intellectual desire for specific knowledge. On the face of it, therefore, the two studies seem to imply that different facets or mechanisms of curiosity may involve (at least partially) separate regions of the brain and may manifest themselves as distinct psychological states.

If confirmed, this interpretation could lend support to Jordan Litman's binary or dual scenario. Recall that Litman proposed the existence of what he dubbed I-curiosity, the pleasurable emotion involved with interest, and D-curiosity, the aversive feeling of deprivation resulting from not having access to certain information. Combining the neuroscientific results with Litman's conception conveys the impression that perceptual curiosity should perhaps be classified principally as D-type, while epistemic curiosity is basically of the I-type. This emerging picture is also consistent with the hypothesis by cognitive scientists Gottlieb, Kidd, and Oudeyer that "rather than using a single optimization process, curiosity is comprised of a *family of mechanisms* that include simple heuristics related to novelty/surprise and measures of learning progress over longer time scales." This does not necessarily mean that different varieties of curiosity employ entirely separate sections of the brain. It could be that different types of curiosity involve some common brain core (such as the regions responsible for the feeling of anticipation) but may also activate somewhat separate circuits and chemicals, even though all of the brain operations have a certain degree of functional connectivity.

Jepma and her colleagues cautiously noted, however, that a few uncertainties that exist in both their and Kang and her collaborators' studies do not permit one to draw definitive conclusions. For instance, because of the fact that in the Kang experiment the trivia questions were always followed by the correct answers, it was not entirely clear whether the activation of particular brain components reflected general anticipation of feedback of some sort, curiosity about the specific correct answer, or a combination of both. This was precisely why Jepma's team sometimes chose not to relieve the uncertainty induced by the blurred images and sometimes showed a totally unrelated clear image. This deliberate differentiation allowed the researchers to separate the activation produced by curiosity about the nature of the object in the image alone from that potentially created by the anticipation of some form of feedback that would perhaps unconfound the blurred pictures.

At the same time, however, Jepma's team acknowledged that the fact that in their own experiment the clear image was revealed in only half of the trials introduced an additional ambiguity into the interpretation of the results. Specifically, it was impossible to determine to what extent the participants experienced uncertainty (and thereby curiosity) about the actual identity of the image, as opposed to uncertainty about whether or not a clear image would eventually be revealed (or a mixture of the two).

These inherent limitations in the Kang and Jepma experiments serve to illustrate how difficult research in cognitive psychology and neuroscience truly is. The brain is such a complex piece of hardware and the mind such a wonderfully elaborate and impenetrable piece of software that even the most carefully planned experiments always leave some room for unpredictability.

Still, I was so impressed with Jepma's experiment that I became extremely curious to know what had led to it and what follow-up, if any, ensued. "Why did you decide to study curiosity?" I asked her in a Skype conversation.

"I was studying the dilemma between exploiting and exploring," she explained. "You exploit things you already know and explore when you know very little. I was interested in how exploitation and exploration guide and direct your decision process."

While this made perfect sense, it was still not a complete answer to my question, so I persisted. "And then?"

"Well, I realized that a chief motivation for exploration is curiosity, so that's how I got into it. To my surprise, I discovered that very little research had been done on curiosity on the neuroscientific side, in spite of its enormous importance."

"Have you done any additional work that perhaps hasn't been published yet?"

She smiled. "How did you guess? I did a preliminary study to test whether individuals are prepared to endure even physical pain to relieve their curiosity."

"And are they?"

"Not all were willing to suffer pain," she said, "but a few were. There was a significant effect."

All I could think of saying was "Wow!"

Another interesting result came out of both neuroimaging studies. The findings suggest not only some tantalizing connections between curiosity, memory, and learning, but also an overlap between the brain circuits of curiosity and reward. As you may recall, the cognitive investigations also implied that the mind produces rewards that assign value to the gathering of information. Over and above that, the fMRI experiments have raised an entire set of new, deeper questions: How exactly does curiosity influence memory? Does working memory capacity influence curiosity? Is the value of the accumulation of information to the reward system the same as that of other valued goods (such as a piece of chocolate, a glass of water, or a drug)? Is the curiosity that drives volitional, active exploration the same as the curiosity that is artificially induced and passively reduced in the neuroscientific experiments?

Curiosity, Reward, and Memory

In some sense, we didn't really need the neuroimaging studies to discover that people learn a subject more effectively when they are curious about it than when they are bored stiff. We have all felt the combination of weariness and fatigue that accompanies being forced to listen to a tedious lecture or to sit between two dull individuals at the dinner table. People find it much easier to learn about topics they are interested in. But does curiosity also affect what we remember? And if so, through which mechanism? These were the questions that neuroscientists Matthias Gruber, Bernard Gelman, and Charan Ranganath of the University of California, Davis set out to answer.

The researchers started on a path similar to that of Kang and her team, by asking students to work their way through a series of trivia questions. The participants were then instructed to rate their confidence in their answers and to indicate their level of curiosity to find out the correct answer for each question. This was, however, where the Gruber study introduced a new twist. The initial process allowed Gruber and his colleagues to create a custom-made list of questions for each student, a list that left out all the questions to which the student already knew the answers. Every one of these lists was composed of questions about which the student had expressed a wide range of intensity of curiosity, from "dying" to know the answer to not caring the least bit.

The researchers then used fMRI to scan the brain of each student while his or her personalized list of questions appeared in sequence on a screen. Following each question there was a quiescent anticipation interval of fourteen seconds, during which a random face flashed on the screen for two seconds. Then the answer to the question appeared, and the process was repeated. After the brain scan session, the subjects were asked to take a surprise memory test for the faces they had been shown during the waiting period, as well as a memory test for the answers to the trivia questions.

In terms of the brain regions that were activated during the expectation for interesting information, Gruber and his collaborators' results were generally consistent with those of Kang and her colleagues. The Gruber study did, however, provide fascinating new clues connecting curiosity to reward and memory. First, by comparing the brain activity between the set of trials in which the subjects were extremely curious to know the answer and that in which they were not, the researchers discovered that the activation precisely followed the pathways in the brain that transmit dopamine signals. Dopamine is a neurotransmitter—a chemical released in the brain by nerve cells to send signals to other nerve cells—that plays an important role in the brain's reward system. Gruber and his colleagues' results therefore confirmed that epistemic curiosity taps into the reward circuitry. In other words, the desire to learn produces its own internal rewards. Second, as could be expected, the study revealed that when people's curiosity was piqued, they learned more readily. They also retained the information better twenty-four hours later. More surprising, however, the study also showed that the subjects were even better at recognizing the random faces that flashed up on the screen while they were waiting for the answers to questions they were curious about. The implication was that even the learning of incidental information was improved in high-curiosity conditions. Gruber speculated, "Curiosity may put the brain in a state that allows it to learn and return any kind of information, like a vortex that sucks in what you are motivated to learn, and also everything around it."

A third finding uncovered by Gruber and his team was equally interesting. They realized that not only was the learning process associated with increased activity in the region that plays a crucial role in the formation of new memories, the hippocampus, but that the strength of the interaction between the hippocampus and the reward circuit was also boosted. It was as if curiosity actively recruited the reward system to assist the hippocampus in absorbing and retaining information.

Experiments by psychologists Brian Anderson and Steven Yantis of Johns Hopkins University added yet another dimension to this picture. They showed that the relation between curiosity and the reward system acted in the opposite direction as well. That is, stimuli that had been previously associated with reward generated curiosity and captured attention more than half a year later, even when the original information had been presented as irrelevant distractors. It therefore appears that stimuli that are initially followed by the delivery of rewards generate persisting attentional biases and induce curiosity, even without continued reinforcement. In other words, the interaction between curiosity and the reward system is a two-way street, with each side assisting the other.

Finally, Gruber's results seem to suggest that even though curiosity reflects an intrinsic motivation, it may still be mediated by mechanisms and brain circuits similar to those that make people yearn for, say, ice cream, nicotine, or winning in poker games. Does this mean, however, that curiosity and the information it seeks only modulate somewhat the value the brain assigns to primary rewards, such as water or food? Or do information and its acquisition have their own independent value somewhere in the brain?

To investigate this question, neuroscientists Tommy Blanchard, Ben Hayden, and Ethan Bromberg-Martin recently used the fact that advance information about future events helps in decision making to test competing hypotheses about where the brain actually evaluates potential rewards. They concentrated on an area in the frontal lobes of the brain of monkeys that is known to be involved in the cognitive process of decision making. Specifically, they recorded the activity of neurons in a region known as area 13 of the orbitofrontal cortex (OFC; Figure 18). The OFC plays a central part in signaling information about reward.

The researchers were attempting to clarify the following point. Whereas there is no doubt that the values assigned by the brain to information and to primary reward (such as food or drugs) are eventually integrated into a single quantity, which in turn is used to guide a particular behavior, it is not known what exactly happens *before* the two values are combined to create one aggregate. The researchers' goal was, therefore, to distinguish between two potential alternatives concerning the role of the OFC in this type of decision making. The first possibility was that the OFC represents a stage at which components such as information and primary reward data are kept completely separate, only to be combined later in some downstream area. Alternatively, the OFC could be precisely the location where information and primary reward factors are already fused together to generate the single value that eventually guides decisions.

In their study, Blanchard and his colleagues recorded the activity of OFC neurons in the brains of monkeys who could choose between gambles that differed in two ways: (1) the amount of water associated with winning the gamble (a primary reward) and (2) the informativeness—whether a cue revealed the gamble's outcome before its delivery.

Two results were particularly important. First, monkeys regularly sacrificed water to gain advance information. This is reminiscent of Jepma's tentative finding that people were prepared even to endure pain to satisfy their curiosity. Second, the OFC was found to encode the information value and the primary reward value independently rather than to integrate them into a single variable. Philosopher Thomas Hobbes was apparently onto something when he referred to curiosity as the "lust of the mind." In fact, Blanchard, Hayden, and Bromberg-Martin speculated that "just as the OFC regulates seeking of appetitive reward in response to internal states like hunger and thirst, the OFC may regulate information seeking in response to internal states like uncertainty and curiosity." Simply put, the OFC appears to serve as a gateway to the rest of the reward system, and it generates inputs that are later used in the consolidated evaluation process, but it does not act as the final evaluator. In particular, curiosity seems to be quantified separately from the other elements that the OFC appraises.

All of these experiments demonstrate that while the curiosity jigsaw puzzle is far from complete, neuroscientists are starting to reveal the intimate connections among the mechanisms of curiosity, reward, and learning, and also to identify the specific roles of various brain constituents in the entangled circuitry of these mechanisms.

Willpower

The procedures adopted in the studies of Kang, Jepma, Gruber, Blanchard, and their collaborators did not allow the researchers to examine the question of whether curiosity relief through passive exposure to uncertainty-reducing information (such as answers to the trivia questions or clear images that demystify blurry ones) differs from curiosity satisfaction achieved through active exploration.

In one of the attempts to close this gap in the understanding of how curiosity works, University of Illinois cognitive neuroscientist Joel Voss and his collaborators investigated what happens in the brain during active exploratory behavior driven by one's own free will.

Voss and his team correctly observed that whereas most theories of learning emphasize the importance of the individual's control over what is being learned and over how and when it is being learned, most past experiments on curiosity and learning used paradigms in which participants passively reacted to information presented to them. To avoid this drawback, Voss and his colleagues used a learning task devised so that they could study the effects of volitional (by choice) control of visual exploration on the efficiency of the learning process. Specifically, participants were asked to examine arrays of common objects, viewing one object at a time through a moving window. So far this sounds pretty conventional, but here is the new angle. Each participant experienced two viewing conditions: one in which the subject could actively control the window's position and another in which he or she was a passive recipient of the string of images. Voss and his team used an ingenious technique, in which the self-controlled, volitional movements of one participant were recorded and then displayed during the passive condition of the next subject. Overall, participants viewed precisely the same sequences, presented with strictly the same time intervals, in both the volitional and passive conditions, but in the first instance the participants chose the sequence they viewed. This method allowed the researchers to identify the differences that could be directly attributed to the effects of volitional control.

The results showed that volitional control significantly enhanced later memory relative to the passive configuration, even though the information content was identical. This will probably not surprise anyone who ever tried to deduce some information from a website while someone else was in control of the computer's mouse.

Perhaps even more important, the activation of the hippocampus, which plays a central role in the consolidation of information from short-term memory to long-term memory, was stronger during volitional, active exploration. The researchers suggested, therefore, that the effects of volitional control on memory could be ascribed to an increased coordination between the hippocampus and other cortical areas in the brain. Recall that Jepma and her colleagues also found that the relief of perceptual curiosity was associated with enhanced hippocampal activation and increased incidental memory. The Voss study augmented this picture and amplified it by indicating that volitional control further strengthens learning. Voss and his collaborators theorized that the additional effect was the result of a significant boosting of the communication between the hippocampus and the neural systems responsible for such functions as planning and attention. This enhanced communication, in turn, produces a more efficient updating process, which allows the brain to be curious about and absorb the most salient features of the available information. This is in some sense the brain's version of our emergency management center, which coordinates communication among agencies that have to respond to disasters.

Before I briefly summarize what we have learned about the nature of curiosity from the cognitive and neuroscientific experiments, I would like to mention two more caveats to be aware of. First, in task-based fMRI experiments, the researchers examine the spatial extent (that is, the locations) of the brain activity at prescribed times. This is tantamount to assuming that the activity takes the form of a *standing wave* (or *stationary* wave; such as the one formed in a vibrating violin string held fixed at both ends), where at each point along the wave the strength of the signal remains constant in time. However, in a study published in June 2015, neuroscientist David Alexander of the University of Leuven in Belgium and his colleagues argued that brain hustle and bustle is more akin to *traveling waves*, where activations and deactivations move rapidly through the brain. This means that treating the temporal and spatial dimensions as if they are distinct may result in the loss of much of the relevant information. Alexander and his collaborators concluded, "We question the very notion that neurological entities are events [that] occur at certain locations and times, rather than being comprised of trajectories that extend over locations and times." In other words, this team argues that just as a photograph of a small portion of the sea does not convey the full picture of a turbulent ocean, by examining what happens only at particular regions of the brain at a fixed time, one misses the fact that the activity propagates in a complex fashion throughout the brain. If Alexander and his team are correct, then some of the conclusions from neuroimaging may have to be revised when more sophisticated imaging and data analysis techniques become available.

A second qualification is related to the confidence we can have in the results of psychological research in general. In an important collaborative study entitled "The Reproducibility Project: Psychology," that was published in August 2015, 270 researchers across five continents reported that they were able to replicate only about forty percent of the results from 100 studies in cognitive and social psychology that had been published in 2008 in respected scientific journals. This project is part of the application of the scientific method, which advocates continuous testing, rechecking, and questioning the validity of hypotheses. It is only by adopting such rigorous scrutiny procedures that science can be self-correcting. While the Reproducibility Project has been to some extent hoisted with its own petard—a more recent study raised questions about the results of the Reproducibility Project itself—it remains true that we should always exercise caution and emphasize uncertainty when evaluating experimental results in general, especially those that supposedly provide empirical evidence for certain theories favored by the experimenters. Note also that because of technical and funding-related difficulties, studies in neuroscience often involve a relatively small number of subjects. For example, the experiments of Kang and Jepma scanned the brains of only nineteen students each. Consequently, the statistical significance of the results is limited.

Bearing these important caveats in mind, what is the picture of curiosity that tentatively emerges from all the recent psychological and neuroscientific studies? Here is a very brief overview.

Moving into Focus

Curiosity has only relatively recently started to receive the focus it deserves. While many of the details of the mechanisms underlying curiosity are still unknown, at least a broad-brush understanding is beginning to emerge. What have we learned so far?

First, as children engage in increasingly complex activities, they explore their new environments and acquire fresh knowledge. The trajectory that they follow while growing up is remarkably similar for most children, which indicates common underlying mechanisms. Children's curiosity seems to drive them along a path that increases knowledge and involves a well-suited decision process that maximizes learning and facilitates the discovery of causal links. Children seem to understand relatively early that every effect is associated with a cause in an unbroken chain of events. Their curiosity appears to assign value to competing tasks based on the potential of these tasks to enable discovery.

The exploratory behavior of adults also seems to follow rather consistent patterns, even in open-ended circumstances and in spite of individual differences. Artificial-intelligence researchers Frederic Kaplan and Pierre-Yves Oudeyer suggested that all of these elements could be captured in the context of a paradigm in which the goal of curiosity and exploratory behavior is to reduce prediction error to the extent possible. In other words, according to this view, humans (children and adults) avoid both extremely predictable and highly unpredictable exploratory routes in order to focus on those courses to satisfying curiosity that can maximize the rate at which their prediction errors decrease. Gottlieb, Kidd, and Oudeyer further clarified and expanded what they perceive as curiosity's main "goal," which they say is to maximize learning (rather than merely to reduce uncertainty).

What is curiosity really? In my humble view, the cognitive and neuroimaging studies appear to support a scenario in which what we refer to as curiosity may actually encompass a family of intertwined states or mechanisms that are powered by distinct circuits in the brain. In particular, the curiosity triggered by novelty, surprise, or puzzling stimuli—perceptual curiosity—seems to be primarily associated with an unpleasant, aversive condition. In this case, curiosity is a means to reduce the negative feeling of deprivation. This type of curiosity is adequately explained by the information-gap theory, and its intensity as a function of the level of uncertainty generally follows an inverted-U shape.

On the other hand, the curiosity that embodies our love of knowledge and the drive for its acquisition—epistemic curiosity—is experienced as a pleasurable state. In this case, curiosity provides an intrinsic motivation for its own sake. Consistent with this picture of different types of curiosity, perceptual curiosity was found to activate brain regions that are sensitive to conflict, while epistemic curiosity switched on brain areas linked to the anticipation of reward.

The satisfaction of curiosity (of any type) is closely related to the neural reward circuit, and it enhances memory and learning, especially when the information violates prior expectations and when the exploration is active and volitional. In the other direction, past rewards can trigger a higher level of curiosity, even without reminders or boosting.

An interesting recent study suggests that even individual differences can be estimated with some confidence using fMRI. Neuroscientists Ido Tavor and Saad Jbabdi of the University of Oxford and their collaborators showed that fMRI imaging of a person's brain while that individual is at rest, doing absolutely nothing, can predict which parts of the brain will be activated during a range of active tasks. Those tasks included reading (which involves language interpretation) and gambling (which is associated with decision making).

As I noted earlier, these new insights do not mean that we now understand curiosity. Curiosity is a topic where ideas jostle each other and everything can and is likely to change. Here are just a few of the basic questions neuroscientists and psychologists would like to have more complete answers to: Does curiosity play a role in maintaining cognitive abilities during adulthood? What are the precise similarities and differences between curiosity and other basic drives, such as hunger, thirst, and sexual desire? What are the main neural elements and mechanisms that govern and direct curiosity? How exactly does the brain merge those components to construct a clear course of decision making? What precisely underlies the individual differences in curiosity and exploratory drives?

These are not easy questions, and considerably more research is required to provide unambiguous answers to all of them. Concerning the last question, for instance, Gottlieb, Kidd, Oudeyer, and their collaborators are embarking on an extensive study aimed at testing the interesting hypothesis that an important component of the variations among individuals' curiosity is related to differences in their working memory capacity and executive control. The researchers speculate that since working memory directly affects the encoding and retention of information, it can impact the value we place on learning and novelty. To assess the viability of their conjecture, the researchers will search for correlations between curiosity and measures of working memory in a group of children. They will first determine a ranking of curiosity among the children based on a number of exploration tasks, then they will characterize the children's working memory capacity with standard memory tests. These experiments (with more than 100 children) will allow the researchers to statistically examine whether curiosity and working memory are indeed correlated. It is interesting to note in this respect that as early as the 1960s, psychologist Sarnoff Mednick suggested that creativity (for which curiosity is a necessary ingredient) is merely an expression of an associative memory, the ability to remember the relationship between unrelated items, that works exceptionally well.

There is another aspect of curiosity that deserves special attention. Humans are different from all other animals in our cognitive ability to formulate and integrate abstract information, in our power to invent and analyze hypothetical and even fictional scenarios, and in our aptitude for turning almost everything we perceive into meaningful *why* and *how* questions. Ultimately, it is this curiosity and desire to explore in order to get to the bottom of causes and effects that has led to the birth of religions, to disciplines such as logic (and thereby mathematics and philosophy), and to the quest to understand how nature works—what we refer to today as science, and subsequently to technology and engineering, since most research eventually leads to applications. At the same time, the appearance and evolution of the remarkably intricate human language, and the inherent mental power to describe what exists not only in the real world but also in a world that can only be imagined, have spawned literature, the visual arts, and music.

When and why did this marked difference between the curiosity manifested by humans and other animals arise? In the next chapter I investigate how our ability to ask "Why?" is a prerequisite for sophisticated forms of curiosity and is uniquely human.

SEVEN

A Brief Account of the Rise of Human Curiosity

MODERN RESEARCH IN PSYCHOLOGY AND NEUroscience suggests that curiosity (at least the epistemic kind) is a mental decision process that aims at maximizing learning. To achieve this goal, it assigns values to competing alternatives based on their perceived potential to answer questions the individual finds intriguing. In essence, therefore, curiosity is really an engine of discovery.

The fMRI studies have allowed researchers to locate curiosity in the brain. They have shown that the main regions of the brain that actively participate in the cognitive processes of arousal and satisfaction of curiosity belong either to the cerebral cortex, the outer layer of neural tissue that is the headquarters of memory, thought, and consciousness (as well as motor and sensory functions), or to the striatum, a subcortical part of the forebrain that is pivotal for the reward system (Figure 18). Consequently, to ask why humans are the only species capable of incessantly asking "Why?" is, at some level, equivalent to wondering what makes the human cerebral cortex and striatum unique among animal species. At the same time, we would also like to understand (from an evolutionary perspective) how those human brain structures became the way they are. Before we begin to answer these questions, however, it may be helpful to review a few simple facts concerning the human brain.

Neurons are the core components, the computational building blocks that create brain activity. These electrically excitable cells are the units that process and transmit information through a variety of chemical and electrical signals. As in a vast computer network, each neuron connects with thousands of its neighbors. The connections occur in two kinds of branches: *axons*, which transmit signals away from the nucleus of the cell, and *dendrites*, which receive incoming signals. There is a tiny gap, a synapse, where an axon meets a dendrite. When a neuron is activated, the axon secretes chemicals known as neurotransmitters into the synapse. This allows the electric signal to cross the gap and cause another neuron to fire up. As in a fast-moving forest fire, many neurons can thus be nearly simultaneously activated through a chain reaction.

The human brain has two hemispheres, which are covered by a deeply wrinkled gray tissue, the cerebral cortex (Figure 18). Each bulging area on the surface is a gyrus, and each infold is a sulcus. The important point for our purposes is that part of the neurons in the cerebral cortex are responsible for everything we associate with the concept of intelligence.

Brainy Matters

Oddly enough, until about 2007, even though sampling methods based on two-dimensional sections of the brain (stereology) had been extensively used, the total (average) number of neurons in the human brain, or in the brain of any other species, for that matter, was not known with great accuracy. While the number 100 billion was often quoted for the human brain, it was not particularly reliable. Similarly the number of neurons in any of the brain's substructures was equally uncertain. All of this changed with the admirable work of Brazilian researcher Suzana Herculano-Houzel and her team. Herculano-Houzel devised an ingenious method for counting neurons by simply dissolving the brain into a "soup"—a suspension of free cell nuclei. Since the soup could be shaken and mixed thoroughly to transform it into a homogeneous solution, counting the neurons in a sample of the liquid and multiplying the number obtained by the appropriate ratio of volumes gave Herculano-Houzel a fairly accurate determination of the number of neurons in the entire brain, and similarly in any of the brain's components.

I first met Herculano-Houzel in 2013 and later talked to her in more detail about her work as I was writing this chapter. In one blow, she and her colleagues put an end to years of ambiguity and speculation and replaced them with solid data. So, you may impatiently wonder, how many neurons are there in the human brain? Herculano-Houzel's answer was unequivocal: on the average, for Brazilian males fifty to seventy years old, about 86 billion. A rat, by comparison, has only about 189 million (that explains why a rat isn't writing this book), and an orangutan has about 30 billion. You may think that 86 billion is sufficiently close to the original estimate of 100 billion and that, therefore, the added precision is not very important. Herculano-Houzel's answer to such comments is to point out that the difference of 14 billion neurons constitutes the entire baboon brain! She and her research team also calculated average numbers of neurons for the human brain's main parts: 69 billion in the cerebellum (the part that is vital for motor control), 16 billion in the cerebral cortex, and slightly fewer than 1 billion in the rest of the brain.

Herculano-Houzel's work, however, supplied much more information than mere neuron counts. It opened the door to a variety of new insights. In particular, Jon Kaas, a neuroscientist at Vanderbilt University, Herculano-Houzel, and their collaborators, were able for the first time to show that not all brains are built by the same scaling rules. In the brains of rodents, for instance, a ten times larger number of neurons in the cerebral cortex requires a cerebral cortex that is not ten but about fifty times larger in mass. In contrast, primates manage to cram more neurons into relatively smaller brains and also into smaller cerebral cortices. In fact, the brain mass in primates scales roughly linearly with the number of neurons; that is, a doubling in brain mass results in doubling the number of neurons. For example, the brain of the rhesus monkey weighs about 87 grams, which is eleven times heavier than the brain of a marmoset, and the rhesus brain has about ten times more neurons than the marmoset brain.

As primates, humans benefited from this more efficient packing of larger numbers of neurons into a smaller mass in the cerebral cortex and the prefrontal cortex. *This neuron jamming has given us humans our first clear evolutionary advantage, at least over nonprimate species.* In fact, a study by German neurobiologists Gerhard Roth and Ursula Dicke showed that intelligence across species is tightly correlated with the number of neurons in the cerebral cortex. This, however, can't be the whole story. You may still wonder why other primates aren't able to ask (and often answer) *why* questions. Even more acutely, why aren't they investigating our brains?

How do we know that chimpanzees don't ask "Why?" There exists a considerable amount of experimental evidence showing that chimpanzees do not seek explanations for forces or causes that are not directly observable in the way humans do. In an interesting experiment by Daniel Povinelli and Sarah Dunphy-Lelii of the University of Louisiana at Lafayette, for example, the researchers designed sham wooden blocks that couldn't be made to stably stand up because of small lead weights placed inside. The sham blocks and visually identical, functional blocks were presented to both three- to five-year-old children and chimpanzees. The results were quite stunning. Sixty-one percent of the children engaged in at least one form of inspection of the bottom of the sham block. Moreover,

50 percent of the children performed both visual and tactile inspections. Not one of the seven chimpanzees engaged in any form of inspection. All seven chimpanzees just kept trying to get the sham block to stand up. They were simply unable to wonder *why?*

A fascinating experiment in 2015 may have identified the specific area of the brain that gives humans their unique ability to process abstract information. A team of researchers led by cognitive neuroscientists Stanislas Dehaene and L. Wang looked at the differences in activation in the brains of humans and macaque monkeys as both listened to a few sequences of tones. The sequences differed in two respects: the total number of tones (investigating the ability to count) and the arrangement of the tones (indicating an ability to recognize abstract patterns). The team used fMRI to monitor the brains as the string of tones changed. Changes could be of the type AAAB being replaced by AAAAB (where the pattern remains constant but the number changes) or AAAB replaced by AAAA (where the number is constant but the pattern changes). Dehaene and his colleagues also examined sequences in which both the number and the pattern changed simultaneously, as in AAAB transforming into AAAAA. In both humans and macaques the area of the brain normally associated with numbers was activated as the number of tones changed. Both species also registered changes in repeating patterns in the corresponding brain areas. However, only human brains registered an additional intense response in the inferior frontal gyrus (the region associated with learning and language comprehension) when both the number and sequence pattern were changed. The implication is that while the monkeys recognize numbers and patterns, they don't find the abstract combination of the two sufficiently interesting to investigate further. These findings may be relevant to other uniquely human characteristics, such as music appreciation.

But why is there such a difference between humans and monkeys? Before exploring this question, I want to investigate another aspect of the human brain that seems rather perplexing, which is related to the brain's energy consumption.

The operation of the human brain costs about 20 to 25 percent of the energy budget of the entire body, in spite of the fact that the mass of the brain is only about 2 percent of the body's total mass. In comparison, the brains of other species are much "cheaper" to run, at an average cost that doesn't typically exceed 10 percent. What is it that makes the human brain's energy bill so excessively high? Herculano-Houzel and her team were able to give a clear answer to this question as well: the human brain consumes (relative to the body) more energy simply because it has a much larger number of neurons than the brain of any other primate. As it turns out, the energy consumption per neuron actually varies relatively little, even across different species. The high metabolic cost of the human brain is nothing more than a direct consequence of the fact that it has such a large number of neurons.

Our brains, like those of other animals, are a natural product of Darwinian evolution. Human brains are expensive to fuel because they contain more neurons for their volume compared to the brains of nonprimates. But this still leaves us with an intriguing question: Why do we have so many neurons while gorillas don't, even though they are primates and they have a much larger body size?

Big or Smart?

Animals in the wild don't have the luxury of going to the nearest supermarket and buying as much food as their credit card allows. (Actually, it's a sad reality that many people don't have that luxury either.) They have to forage for their meals. There is a limit, however, to how many hours per day they can spend on searching, hunting, chewing, and eating before their health starts to deteriorate, since they also need to sleep, care for their young, and avoid predators. That limit is typically not higher than about eight to nine hours. This means that, on average, any given animal, primates included, cannot expect to gain more than a certain amount of energy from food each day. From extensive observations of a number of species in the wild, researchers have concluded that for primates, the daily intake depends on the primate's mass in such a way that a species that is 10 times heavier than another can accumulate and eat (for the same daily duration of foraging) about 3.4 times more calories than the smaller species.

At the same time they procure energy, however, the various species also consume energy, both for the operation of their body and for the functioning of the neurons in their brain. And this is where the limit appears. First, it turns out that the rate of physical (bodily) energy consumption is a steeper function of body weight than the rate of energy accretion through foraging. Quantitatively, the metabolic cost of the body of a species that is 10 times heavier is about 5.6 times higher (compared to only a 3.4 times higher energy input for the same time spent on foraging). This in itself limits the body size a primate can have when it spends the maximum possible time foraging. Herculano-Houzel and her colleagues calculated that top weight to be around 265 pounds—close to the weight of a silverback gorilla that is not an alpha male (not the leader of the pack).

The situation becomes even more intriguing when we throw into the mix the additional caloric cost of large numbers of neurons in the brain. In fact, it immediately becomes clear that even if primates spend the maximum time foraging that their physiology allows them to sustain in the long run (about eight to nine hours), they cannot afford both a large body and a large number of neurons. As Herculano-Houzel puts it, "It is brains or brawn." One of those has to come at the expense of the other. To be more specific, researchers estimate that even if primates in the wild were to spend the full eight hours feeding every day, the maximal number of neurons they would be able to support would be about 53 billion (still far fewer than a human's 86 billion). Even that number, however, would come at the price of the body mass not exceeding about fifty-five pounds! Trading brain power for body weight (if evolution had allowed for such choices), a primate weighing 165 pounds would flaunt only about 30 billion neurons—about one-third the number in the human brain (Figure 20 shows the brain mass and

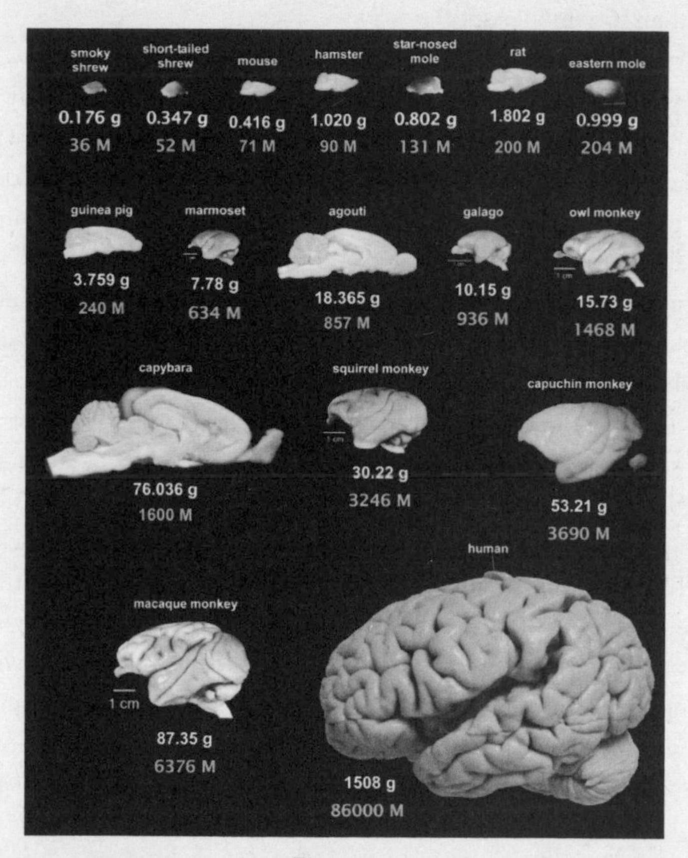

Figure 20

total number of neurons for a few mammalian species). This seems to have been roughly the number of neurons boasted about 6 million years ago by the last common ancestor of today's chimpanzees and humans. Then, from about 4.5 million years ago, the number of fossils of hominins was found to increase dramatically. One such discovery became especially famous; the fossil skeleton of a nearly human female dated to be 3.2 million years old (Figure 21 shows a cast from the Muséum National d'Histoire Naturelle, Paris) demonstrated the clear divergence between the human ancestors and the lineage that has led to modern chimpanzees and bonobos.

Paleoanthropologist Donald Johanson discovered this skeleton, known as "Lucy," in Hadar, in northern Ethiopia, on November

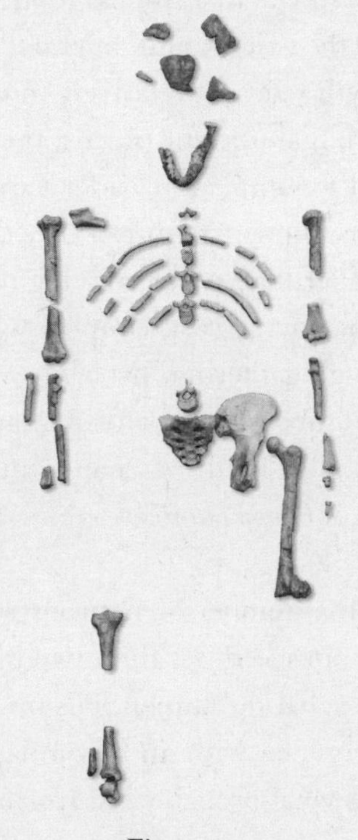

Figure 21

24, 1974. The name, by the way, was suggested by expedition member Pamela Alderman, who was inspired by the Beatles song "Lucy in the Sky with Diamonds." Lucy's skeleton, as well as a collection of scattered remains of at least thirteen other individuals discovered in Hadar in 1975 and a bone found in 2011, are thought to represent members of the hominin species *Australopithecus afarensis*. From the structure of the foot, the knee, and the spine, paleoanthropologists have concluded that Lucy was about three and a half feet tall and walked mostly upright. In terms of her diet, she was a vegetarian, who, like modern chimpanzees, ate mainly fruits.

If the clear separation of Lucy's *Australopithecus* (meaning "southern ape") genus from the ancestors of some modern great apes wasn't surprising enough, what came later was absolutely astounding: the brain of the hominin species that has led to modern humans almost tripled in size within the past 1.5 million years!

At first, the growth rate was relatively modest. Once Lucy and her kind became habitual upright bipeds, they were able to cover longer distances and to sample a broader variety of environments since the caloric consumption required to walk on two feet is almost four times smaller than when moving on feet and knuckles. This reduction in the energy cost, coupled with access to a wider range of foods through gathering, probably allowed for a moderate increase in the number of neurons in a later species known as *Homo habilis* (the "skilled man" or "handy man") about 2 million years ago. The brain of *Homo habilis* was already larger than that of modern gorillas.

The increase in the number of neurons and in brain capacity started to really pick up speed less than two million years ago. It is tempting to speculate that the stupendous upsurge in brain power was intimately intertwined with an accompanying rise in human curiosity. Curiosity is what probably led *Homo habilis* to the invention of the first tools—sharp-edged stones formed by banging two rocks against each other. Once those tools were manufactured, curiosity again helped *Homo habilis* in recognizing them as a solution to two problems Lucy and her cohort couldn't easily resolve: separating meat from bones and cutting it into more easily digestible pieces, and extracting bone marrow from the bones of carcasses. As their own teeth and the skeletal remains of what constituted their meals indicate, members of the *Homo habilis* species significantly enriched their caloric intake by moving away from strict vegetarianism and incorporating meat as a regular component of their food.

The next major step along the way to modern humans dates to about 1.8 million years ago, with the species known as *Homo erectus.* Members of this long-legged, short-toed species were probably excellent endurance runners—a characteristic that must have helped them in hunting live (although at first small) animals rather than just scavenging cadavers.

All of these new, improved features contributed, no doubt, to the buildup of the number of neurons in the brains of the *Homo* species. Natural selection pressure most probably also played a part, since the organization and execution of hunting expeditions must have required enhanced cognitive abilities compared to those needed for the mere digging out of plant roots. The key question still remains, however: What enabled the more than doubling in size of the hominin brain between *Homo erectus* and *Homo sapiens*? That phenomenal change occured within less than a million years. As we shall see, it may have been due to something that today we take for granted.

Food for Thought?

The energy limitations imposed on the number of affordable neurons were very real. In order to somehow circumvent those, members of the *Homo erectus* and even more so of the archaic *Homo heidelbergensis* species had to find a way to significantly magnify the effectiveness of their caloric intake. Fortunately for our ancestors and us, one of the best ways to do that is by cooking. In addition to improved taste (unless the cook happens to be really terrible), cooking permits a much more efficient digestion because it breaks down food into much smaller scales, both on the macroscopic level (by mincing and mashing) and on the molecular level (by heating), thereby exposing it to enzymes in the digestive system. Cooking gelatinizes the matrix of collagen in animal flesh and denatures complex molecules in plants. Furthermore, the advent of cooking added to the human menu an entire series of foods (for instance cereals and rice) that were previously indigestible.

In his 2009 book, *Catching Fire*, Harvard primatologist Richard Wrangham speculated that the introduction of cooked meat into *Homo*'s diet directly influenced the evolution of the human brain. By examining the rigid energy constraints on the number of neurons, and not just the brain size, Herculano-Houzel turned this hunch, according to which we have to thank cooking for the large number of neurons in our brains, into an even more plausible hypothesis.

What I find especially fascinating is that if Wrangham's and Herculano-Houzel's suggestion is correct, *then curiosity may have been instrumental in the rapid rise in the number of neurons* through a mechanism of positive feedback amplification.

Here is how this scenario might have worked. Curiosity was surely behind the fact that members of the *Homo* species (probably *Homo erectus*) discovered that fire could be helpful and at some point started to use it as part of their way of life. Fire was essential for more than cooking; it provided heat and light and allowed humans to migrate to areas at higher latitudes. The earliest evidence of what appears to be controlled use of fire is about 1.6 million years old, and it comes from two sites in Kenya, Koobi Fora and Chesowanja. Burned bones and plants that are about a million years old were also found at the Wonderwerk Cave, near the edge of the Kalahari Desert in South Africa. Similarly, burned flints and wood found in a hearth-like pattern in Israel date to about 790,000 years ago. The emergence of constant, habitual fire use probably came later; clear signs were found in the Tabun Cave in Israel, dating to 350,000 years ago, with similar findings at Schöningen in Germany. In the summer of 2016, archaeologists discovered evidence for the consumption of cooked meat in the Qesem Cave in Israel, a site that is approximately 400,000 years old. Curiosity probably also played a part in the discovery that cooking can soften raw food, make it easier to digest, and make it taste better. Evidence from the shape of the skull dating to this period shows that the facial muscles used for chewing, as well as the teeth, had decreased in size. This is not surprising, given that cooking may have reduced the time spent on chewing from about five hours per day to just one. Evolution also led to smaller gastrointestinal organs through an improved diet, in turn saving expensive digestive energy—basically trading guts for brains.

All of these changes allowed the *Homo* lineage to overcome the energy limitations on the number of neurons, eventually yielding a brain doubled in size. Simultaneously, it was probably the enormous increase in the number of neurons in the cerebral cortex (and the less dramatic but still impressive rise in the number of neurons in the striatum) that presumably brought human curiosity to the point where humans gained a qualitative advantage over other primates. Maybe *Homo* individuals still did not have the *ability* to start asking the *how* and *why* questions, but that *capacity* was starting to evolve. Once those crucial and seminal questions emerged (perhaps following the inception of human language, as I briefly describe in the next section), there was no stopping humans in their ability to discover and create even more food resources, to establish communities, and eventually to bring the concept of culture into being. All ballooned exponentially. The neuron-rich brain, with its newly elevated curiosity, evolved into an even larger, more intellectually flexible and richer brain.

I should note that not all researchers agree with the speculation that cooking played a dominant role in the brain development of *Homo erectus* or later species. Neurobiologist John Allman of Caltech and paleoanthropologist C. Loring Brace of the University of Michigan, for instance, think that cooking was important only in the past half-million years or less (which hypothesis may be supported by the archaeological evidence for the habitual use of fire). Paleoanthropologist Leslie Aiello of the Wenner-Gren Foundation in New York notes that there is no doubt that several converging factors enforced each other through a feedback loop. These could have included a diet rich in meat, shortened guts, cooking, and upright walking. The precise order in which these energy-saving adaptations occurred is still being debated. However, as I said, I do believe that a *qualitative* change in the nature of curiosity provided an additional crucial ingredient in the stew.

A Few "Curiosity Revolutions"

University of Oxford evolutionary psychologist Robin Dunbar begins his book *Human Evolution* with "The story of human evolution has fascinated us like no other: we seem to have an insatiable curiosity about who we are and where we have come from." Origins have indeed always stimulated curiosity. We strive to understand the origins of our species, of the Earth, and of the universe.

The dramatic increase in the number of neurons endowed *Homo sapiens* with new cognitive capabilities. In particular, it enabled novel mechanisms for information processing, for learning, and for communication. Eventually, our newly acquired mental apparatus led to the emergence of what we identify as the unique human language, probably sometime between 500,000 and 200,000 years ago. Opinions vary as to whether language appeared on the scene through a long, evolutionary, Darwinian-like process or a rather sudden mutation that instilled the language faculty in the human brain in a phase-transition-like fashion (as in water turning into ice). This debate, while truly fascinating in its own right, is largely beyond the scope of our concerns here. As an amusing aside, let me note that the Linguistic Society of Paris banned research on the origin of language in 1866 precisely because it deemed the problem unsolvable by rigorous scientific methods. This ban reflected the challenging reality that unlike, say, the use of fire, it is virtually impossible to track the development of language by following archaeological relics. I find this 1866 ban amusing because researchers still cannot agree on the theory for the origin of the human language, seemingly confirming the Linguistic Society's prescient fears.

From the perspective of the present book, the important point is that it appears quite likely that the emergence of the unique human curiosity and emergence of the distinct human language were strongly correlated. Dunbar suggests that the early purpose of a complex vocal language (as opposed to simple sounds) was none other than gossip! That is, rather than merely conveying very rudimentary information, such as "A pack of wolves is approaching," language was used in larger social groups for descriptive narration, addressing issues beyond the imminently existential but still important for survival. As psychologist Elizabeth Spelke put it, "We can use it [language] to combine anything with anything." While there is no consensus on the validity of Dunbar's theory, it hints at an intimate potential connection between curiosity—a key source for gossip—and language. Other theories suggest that language may have evolved for the exchange of different types of social knowledge (such as symbolic social contracts to guarantee the paternity of one's children). Those theories also contain an important curiosity component. The influential linguist Noam Chomsky does not see language primarily as a means of communication. He contends that "language evolved, and is designed, primarily as an instrument of thought." In that respect it is interesting to note that in 2016 researchers at the University of California, Berkeley, managed to create an "atlas of the brain" that shows how the meanings of different words are distributed across various regions of the organ.

American anthropologist Roy Rappaport, British anthropologist Camilla Power, and others argue that language is only one aspect of something much broader: human symbolic culture. They point out that language works only if a structure of cultural facts has already been established. According to this theory, the appearance of language accompanied some ritual practices. When did those start? The earliest potential evidence for symbolic customs is the use of red ochre pigments in places such as the Blombos Cave in South Africa. That ochre-processing "workshop" is about 100,000 years old. The time coincidence between the fossils of modern humans and symbolic artifacts has convinced a few archaeologists (but not all) that modern anatomy and behavior may have coevolved.

Again, the crucial factor from our vantage point is the fact that socially shared myths, rituals, and symbolism were most likely the first sophisticated responses to nagging *why* and *how* questions and were therefore the fruits of curiosity. The same is true for the conjuring of metaphors, and essentially the entire process of abstract thinking (or Chomsky's "instrument of thought") that has underpinned the emergence of all culture. The chain reaction that resulted from the positive feedback between curiosity and language turned *Homo sapiens* into a powerful intellect, with self-awareness and an inner life. The ability for creative thinking, which is largely powered by curiosity, coupled with the aptness to share accumulated knowledge and to pool intelligence with others, eventually led to a few spectacular developments in the history of humankind. One was the so-called First Agricultural Revolution, the transition from hunting and gathering to a diet of cultivated foods produced by settled agriculture. This Neolithic demographic transition started about 12,500 years ago and involved the domestication of various types of plants and of animals such as dogs, cows, and sheep. Another revolution, some 12,000 years later, consisted of the emergence of a dramatically new perspective on the nature of science: the celebrated scientific revolution that started in Europe toward the end of the Renaissance and persisted till the late eighteenth century.

What epitomized the scientific revolution was the transition from the culture of dogmatic assurance that dominated thought during the Middle Ages to the culture of curiosity that put empirical observation and exploration first. Empiricists such as John Locke and David Hume elevated the evidence and impressions of one's own eyes, and encyclopedists such as Denis Diderot attempted to collect all knowledge into coherent texts. The monumental observational and experimental breakthroughs and avalanche of theoretical concepts produced by individuals such as Copernicus, Galileo, Descartes, Bacon, Newton, Vesalius, Harvey, and others came out of the recognition that humans did not know everything—that both the microcosm and the macrocosm had yet to be thoroughly explored. In fact, all the scientific progress that we witness today is a direct extension of these revolutionary ideas. It is not an accident that NASA decided to name the rover that was launched to explore the Martian surface *Curiosity.*

The simple act of listing the names of a few of the trailblazers of the scientific revolution has led me to the next step in my humble attempt to understand curiosity. Since I cannot interview those great thinkers of the past, I decided to briefly interview a few present-day people who are known to be uncommonly curious. The questions that intrigued me in particular were these: How do exceptionally curious individuals describe and explain their own curiosity? How did they choose what to be curious about?

EIGHT

Curious Minds

EINSTEIN ONCE SAID, "THE IMPORTANT THING is to not stop questioning. Curiosity has its own reason for existing. One cannot help but be in awe when one contemplates the mysteries of eternity, of life, of the marvelous structure of reality. It is enough if one tries merely to comprehend a little of this mystery every day." Some people seem to have followed this advice to the letter—they have been endlessly curious. A few of those people became prominent scientists, writers, engineers, educators, or artists. But most humans are curious, and often it is not about transformative issues but about the tittle-tattle in their lives. In our age of narrowly focused specialization, the polymath—a person with wide-ranging knowledge and interests—has become an endangered species. Still, people with a burning passion for exploration and investigation do exist. One person who stands out for his curiosity, even among illustrious scientists, is physicist Freeman Dyson.

Dyson is credited with having successfully unified different versions of the quantum theory of electromagnetism and light, known as quantum electrodynamics or QED (one of those versions, by the way, was created by Richard Feynman). Following this remarkable achievement, Cornell University made Dyson a full professor without bothering about the fact that he did not even have a PhD. As important as QED is, however, it doesn't even begin to cover the full sweep of Dyson's accomplishments. During his long career, he has worked on an astonishing array of topics, including mathematics, nuclear reactors that produce medical isotopes for hospitals, the magnetic properties of matter, solid-state physics, spaceships propelled by nuclear bombs, astrophysics, biology, and natural theology. He has also been a regular contributor of essays to the *New York Review of Books* and wrote a science fiction story at the age of nine.

Over the years, I have met Dyson on several occasions and always enjoyed having stimulating conversations with him. In the summer of 2014, I finally got around to asking him about his exceptional curiosity. At the age of ninety, he was as sharp as ever.

I started with the obvious. "Were you always curious?"

"I was always asking questions as a child," Dyson answered, "but I did not feel that there was anything unusual about it." This was clearly an understatement. In high school he had already started to think about problems that would later lead to interesting contributions to the mathematical branch of number theory.

"And during your adult life, were you interested in certain things more than in others?"

He thought for a few seconds and then replied, "Mostly I was interested in questions because some friends were working on them. I talked a lot to other people and was intrigued by what they were doing. For instance, I talked to Leslie Orgel [a famous British chemist] about the origin of life, and so I started working on that."

"But, was there any pattern to your curiosity?"

Dyson reflected again, then explained, "I was definitely more interested in the details than in the big picture—in the animals rather than in the zoo. In your area, for instance [referring to astronomy and astrophysics], I worked more in astronomy [the study of particular astrophysical objects] than in cosmology [the study of the universe as a whole]."

"And how did you determine when to move on into a new subject and start a new exploration?"

Dyson laughed. "I have a very short attention span. I tend to give up after two to three weeks. I either solve the problem or altogether leave it."

Wow! I thought to myself. *Just like Leonardo.*

As if reading my mind, Dyson went on to say, "I always thought that being a scientist gives you the 'license' to work on *any* scientific problem. You have to have the willingness to give up on 'normal' interests in order to look into something else."

I continued my previous train of thought and said to myself, *and like Feynman, too.* Finally, I asked Dyson whether he had noticed any clear correlation between curiosity and other personality traits. He replied that he hadn't seen any. I suspect that a few of his scientist colleagues might disagree with this last statement, at least when it concerns Dyson himself. Neurologist and author Oliver Sacks (who sadly passed away while I was writing this book) described Dyson as "subversive" in his scientific creativity: "He feels it's rather important not only to be not orthodox, but to be subversive, and he has done that all his life." In fact, Dyson himself wrote in his 2006 anthology of essays, *The Scientist as Rebel*, "We should try to introduce our children to science today as a rebellion against poverty and ugliness and militarism and economic injustice."

The second extraordinarily curious person I talked to was astronaut and polymath Story Musgrave. I first met Musgrave in 1993, when a team of astronauts was preparing for the first servicing mission to the Hubble Space Telescope. I was an astrophysicist working with Hubble at the time.

As you may recall, shortly after the telescope's launch, NASA discovered to its great dismay that Hubble's primary mirror had been perfectly polished, but to the wrong specifications—a deficit known as "spherical aberration." The outer edge of the mirror was ground too flat. Not by much—roughly by one-fiftieth the thickness of a human hair—but that was enough to make the images rather fuzzy. The astronomical community was in a state of shock and the media were only too happy to latch onto the fact that *Hubble* rhymes with *trouble.* Scientists and engineers worked around the clock to develop a plan that would restore Hubble to its originally anticipated performance. Eventually, researchers charted an ambitious scheme to correct the telescope's blurred vision.

It was the job of a team of seven space shuttle astronauts to rendezvous with the telescope, and in the course of five amazing spacewalks, to install inside Hubble its "eyeglasses"—the corrective optics and a new internally corrected camera. Story Musgrave performed three of those jaw-dropping spacewalks. That would have been enough in the category of personal exploration for most people, but not for Musgrave. He has also completed a Bachelor of Science degree in mathematics and statistics, an MBA degree in operations analysis and computer programming, a BA in chemistry, an MD degree (he worked, part time, as a trauma and emergency room physician), an MS degree in physiology/biophysics, and, for good measure, he wrapped up a master's degree in literature. Oh, and he is also a jet pilot, is interested in photography and industrial design, and has seven children.

I talked to Musgrave again in August 2014 and asked the predictable question: "Why did you study for all of those degrees?" Musgrave did not hesitate: "My curiosity is related to a certain restlessness with not being entirely happy with things the way they are. So I always felt that I need to do something. I always had this energy to explore further."

"Okay, but how did you choose these particular topics?"

"One thing naturally led to another. I started with mathematics and statistical tools, in order to be able to predict for what values of the variables I would get a desirable outcome, when dealing with complex systems." He paused for a second. "Those were the early days of computers, so from mathematics I easily gravitated towards programming and operations analysis. Having seen how computers operate, I became curious about how the brain works. This led to studying chemistry, biophysics, and to medical school. Once I acquired some knowledge about the human body and its limitations, the road was paved to the space program."

I had to admit that, when presented in this way, it all made perfect sense. Still, most of us do not pursue our interests with such vigor and persistence.

Musgrave continued, "All the topics I studied were linked or related." Then, after a brief pause, he added, "Every two- to three-year-old child is curious. The question really is what happens when you come out of that. It seems that in many cases adolescence destroys curiosity."

I have heard this comment made by many people. However, the impression I got from the actual psychological research was that it is only the perceptual (and perhaps diversive) aspects of curiosity (novelty seeking in particular) that decline in the transition into adulthood. Specific and epistemic curiosity—the thirst for knowledge—apparently stay fairly constant throughout much of adult life.

Even before my conversation with Musgrave, I had a brief email exchange with another known polymath, Noam Chomsky. Chomsky is a linguist, cognitive scientist, philosopher, political commentator, and activist who has written more than 100 books. He is one of the most cited scholars of the twentieth century, and his work has been extremely influential in fields ranging from linguistics, psychology, and artificial intelligence to logic, political science, and music theory.

"Interesting topic," Chomsky wrote to me when I told him I was writing about curiosity. When I asked him about the types of questions that make him curious, he wittily replied, "Well, I guess one illustration is that I'm curious as to why you're interested in curiosity."

I didn't give up and sent another email: "What was it that attracted you to the particular topics you are interested in?"

He promptly emailed an answer I found fascinating: "Recognition that language is the most distinctive human capacity, and the core of our mental nature, and that every aspect of it poses great mysteries." I had to agree. Even the extremely brief description of the language evolution/revolution that I recounted in chapter 7 highlights the indispensable role of language in the emergence of modern humans as a species endowed with unique abilities.

Something else occurred to me as I was reading Chomsky's note. If I were to replace the word *language* in his answer with the phrase *the ability to ask "Why?,"* that would perfectly describe why I am interested in curiosity.

You may recall that in the neuroimaging experiments, when subjects were given the correct answers to trivia questions, their inferior frontal gyrus (IFG) lit up. Well, among other components, the IFG in humans contains Broca's Area, an important region for language processing and comprehension. Also, Stanislas Dehaene and his colleagues have tentatively identified the IFG as the brain area that allows humans to analyze abstract information. Language, epistemic curiosity, and the processing of abstract concepts undoubtedly are the essence of what Chomsky calls "the core of our mental nature."

The next curious person I talked to had a very unusual career. Even though in high school Fabiola Gianotti concentrated primarily on literature and music, and the first college degree she obtained was in music (as a pianist), she ended up leading a team of about 3,000 physicists that in 2012 discovered what has been dubbed the "God particle"—the Higgs boson. On January 1, 2016, she became Director-General of the European Organization for Nuclear Research (CERN), which operates the world's largest particle accelerator, the Large Hadron Collider, near Geneva, Switzerland.

"Why did you decide to switch from studying the humanities to studying physics?" I asked Gianotti.

"I was always a curious child," she answered, "always had many questions. At one point I decided that physics will actually allow me to try to *answer* some of those questions."

"It must have been hard though, not having the necessary background?"

"Indeed," she admitted. "At the university I had to adjust at first, from an education in the humanities to the ability to comprehend and address the problems physics was posing."

"But you did maintain your love for music?"

"Absolutely. Music is fundamental to me; I listen to music all the time. I have less time to play now, but I still do it sometimes."

"Do you have any passions other than for physics and music?"

She laughed. "Cooking! I see many similarities between physics and music, and physics and cooking. First of all, elegance is a common theme to physical theories, music, and indeed the ballet, which I used to dream about as a young girl."

"I couldn't agree more," I said.

"Then, in both cooking and physics," Gianotti continued, "you need some rules or laws, but you also need creativity." Unfortunately, I couldn't comment on that, since I have hardly ever cooked. I nevertheless reminded myself that cooking may have played an important role in ensuring that humans have a large number of neurons in their cerebral cortex.

There was another question I felt I had to ask, since it was related to the very risky nature of curiosity as a driver of basic research. The discovery of the Higgs particle marked an incredible success for Gianotti and her team; the search for this hard-to-find particle had lasted about four decades. Yet there is a serious possibility that the Large Hadron Collider (Figure 22) will not discover any other new particles. Given the multibillion-dollar cost of the facility, this could pose a considerable public relations challenge to the new Director-General. "What if you don't find anything else?" I asked.

"In fundamental research there are surprises," she replied. "Sometimes in finding something, and sometimes in not finding it. This is part of the game." Then she added, "Negative results are also important, because they help eliminate certain theories and constrain others."

"It would still be somewhat disappointing," I cautiously remarked.

She agreed. "We would still have to combine all possible approaches, from accelerators, from the experimental searches for the particles that constitute dark matter [matter that doesn't emit light, the existence of which is inferred from astronomical obser-

Figure 22

vations that detect its gravitational influence], and from astrophysics." Intriguingly, about three months after my conversation with Gianotti, two experiments at the Large Hadron Collider revealed hints of the potential existence of a new particle, about 800 times heavier than the proton. Unfortunately, with the accumulation of more data, by the summer of 2016 those hints were shown to be nothing more than a fleeting statistical fluke.

I hesitated to bring up another controversial topic—the *multiverse*. The relatively low value discovered for the mass of the Higgs boson coupled with the possibility that the Large Hadron Collider will not uncover any additional new particles have strengthened a speculative view claiming that our universe is but one member of a huge ensemble of universes. According to this scenario, we shouldn't be surprised by any value for the mass of the Higgs, since in the multiverse, even values previously considered unlikely may be represented in some members of the ensemble. After some hesitation I inquired, "How do you feel about the multiverse idea?"

"Psychologically, I feel that relying on the multiverse as an explanation is a bit like giving up," Gianotti replied. "As an experimental physicist I would like to continue to explore all possibilities."

I was thinking to myself that Jacqueline Gottlieb's psychological experiments (described in chapter 5) have demonstrated that this was the attitude of most curious people—to explore all options. Consequently I felt compelled to ask, "Are you as curious today as you were when you were a child?"

Gianotti did not hesitate: "If anything, I am even more curious. I am driven by curiosity and the pleasure of learning. Nothing delights me more than understanding something I didn't understand before." These were almost precisely the words used by Gottlieb, who said, "My greatest joy is when I learn something new."

"Do you see any other characteristic that is common to very curious people?"

"Yes," she said, "the ability to think beyond what is known, beyond what is accepted, beyond what is regarded as established."

"Do you think that this applies to curious artists as well?"

"Absolutely. Artists who are curious explore new avenues. They see reality with different eyes. They also go beyond what we see superficially."

"Who are the artists you like best?"

"In music, my favorite is Schubert, because I see him as the most romantic of the composers of the classical period and the most classical among the composers of the romantic period. In the visual arts I particularly like the artists of the Italian Renaissance."

I happened to know that Gianotti's brother, Claudio, was once quoted as having said that Fabiola "never left anything half-done." I therefore couldn't resist making the following last remark, half-jokingly: "In spite of being extraordinarily curious, as was Leonardo, you actually like to complete your projects."

She laughed. "I will not even begin to compare myself to Leonardo. I really dislike leaving things unfinished. Even if I read a book and I find it to be not very interesting, I would still finish it."

The next person I talked to was someone I have known and admired throughout my entire professional career, beginning with my graduate student days. Martin Rees is a world-renowned cosmologist and astrophysicist and recipient of (among other awards) the Crafoord Prize in Astronomy. He has been Astronomer Royal for the United Kingdom since 1995, was Master of Trinity College in Cambridge from 2004 to 2012, president of the Royal Society between 2005 and 2010, and in 2005 he became Baron of Ludlow. He is one of very few astrophysicists who know just about everything there is to know in astrophysics and cosmology.

In addition to his many accomplishments in astrophysics, Rees has extensively written and spoken about the challenges and risks that humanity is facing in the twenty-first century and about the social, ethical, and political aspects of science. As part of this activity, he cofounded the Centre for the Study of Existential Risk, a research institute at the University of Cambridge that studies potential threats (mostly posed by technology) to humanity's existence.

I started our conversation with the standard question: "Were you extremely curious as a child?"

Rees thought for a few seconds. "I'm not sure," he began. "I do remember having been puzzled by various phenomena. For example, we used to go for holidays to north Wales, and I was interested in the tides. I tried to understand why they were at different times in different places." After a brief pause, he remembered another fact that had perplexed him: "why the tea leaves piled up at the center and bottom of the cup when the tea was being stirred." (This phenomenon is sometimes called the "tea leaf paradox.") Since this interview was at the same time an informal conversation between two scientific colleagues, we couldn't resist the temptation of following Rees's last comment with a brief exchange about hydrodynamics, Ekman layers, and a few other physical concepts. Rees finally returned to the original question: "I was also always attracted to numbers."

I moved on to my second question: "When and why did you decide to go into astrophysics?"

"It wasn't an early decision," Rees recalled. "In the last two years of high school I did specialize in math and physics." Laughing, he added, "Mainly because I wasn't very good at languages." He continued, "I studied math in Cambridge but decided I wasn't cut out to be a mathematician. I did think about doing economics, so I studied some statistics, but in the fourth year I took a few courses in theoretical physics, and then I decided to go into physics. What helped was the fact that I had been assigned Professor

Dennis Sciama as an advisor, and he, who by the way had also been Stephen Hawking's mentor, was a great coach. He created a 'buzz' that swept me along. Consequently, after a year, I was confident that I would do astrophysics."

I wholeheartedly agreed with Rees's appreciation of Sciama, whom I had the honor of knowing personally. Sciama had an infectious enthusiasm for research, an extremely broad knowledge, and an excellent nose for what was worth being curious about in cosmology and astrophysics. I also understood Rees's choice because I found it often to be the case that bright students choose their topics more on the basis of the quality of their teachers than on the intrinsic characteristics of the subjects themselves.

I asked another question: "In recent years, you became more interested in climate change and other existential threats. What inspired these new interests?"

Rees had expected the question and answered it right away: "For a long time I had some interest in politics, and I came to admire socially conscious individuals. Consequently, I became curious about social issues. In my book *Our Final Hour* I made a few points about what I saw as risks, and I believe that those are now generally accepted. Also, when I reached the age of sixty, I was trying to determine what I should do in the following decade, so as to not end up doing nothing. As it turned out," he laughed, "I have been elected to several important positions [alluding to the presidency of the Royal Society and becoming a peer], which gave me the opportunity to be even more involved than I had originally planned."

I decided to include yet another direction of Rees's interests. "Unlike a few of your scientist colleagues, you exhibit more curiosity and tolerance toward theology and religion. Can you briefly describe your views on these topics?"

"I always had an interest in philosophy, and also tolerance towards religion. I am not a religious person myself, but I do appreciate cultural, historical, and religious customs, like going to church on Sundays in Christianity or lighting the Shabbat candles in Judaism, and I would like those to be preserved. I also think that mainstream religion can help in the fight against extreme fundamentalism."

I returned to specific questions about curiosity: "From your experience, do you think some people are much more curious than others, or are different individuals just interested in different things?"

Rees thought for a few seconds. "There surely are various levels of curiosity, but it is also definitely true that different people are interested in different things," he finally replied. "For example, young kids are often interested in dinosaurs and outer space, so the idea would be to start with those topics rather than to force them to become interested in something else." I thought this was excellent advice—to attempt to follow and encourage the curiosity that is already there (at least initially) rather than to impose unattractive subjects.

I happened to know that Rees belongs to a group of futurists who speculate that artificial intelligence may become the dominant species in the not-too-distant future, so I felt that I had to ask about that. "Do you think that 'intelligent' machines will be curious? After all, they may not experience the same types of natural selection pressures that biological life had to evolve through."

Again Rees thought about the question a bit, and finally replied, "The key question is whether they'll have consciousness and self-awareness like us, or whether they'll be more like 'zombies' [a term used to describe machines that are indistinguishable from humans but that lack conscious experience]. If consciousness is an emergent property of complex systems, then they may even have it at a deeper level than us."

"Indeed," I agreed, "but will they be curious?"

"I suppose that depends on how broadly you define curiosity," Rees said after another moment's reflection. "If a mathematician who has relatively little interest in the world outside of mathematics could be called curious, then the machines may definitely have that."

That made perfect sense to me. I finished with my routine question: "Are there any other characteristics that you have noticed which are common to people who appear to be very curious?"

"I am not sure," Rees replied, but then added, "They are generally more intellectually energetic than others. Many of them retain the intellectual playfulness of children—they stay enthusiastic."

This was an interesting way of putting it. Maybe people who are extraordinarily curious are able to keep possession of their perceptual curiosity longer—their ability to be constantly surprised—whereas this quality tends to decay with age for others.

If you thought Gianotti had an unusual career, consider the next person I talked to. Brian May was the famously poodle-haired lead guitarist for the rock band Queen and the composer of such mega-hits as "We Will Rock You," "I Want It All," "Who Wants to Live Forever," and "The Show Must Go On." Believe it or not, he also holds a PhD in astrophysics from Imperial College in London; he was the chancellor of Liverpool John Moores University from 2008 to 2013; he is a science team collaborator on NASA's New Horizons mission to Pluto; he is an expert on and collector of Victorian stereophotography, a technique in which two flat images are fused with a special viewer to produce a 3D scene; and he is a passionate activist promoting animal welfare. No wonder, then, that I wanted to talk to him—very few people today display such a wide range of interests.

I knew that, at age sixteen, May had designed and built his famous guitar, "Red Special," with help from his father. They used the wood from a century-old fireplace mantel to construct the guitar's neck. My first question was therefore, "Why did you decide to build the guitar rather than buy one?"

May laughed. "The simple answer is that we didn't have the money. This was the time of the birth of rock 'n' roll, and the known American guitars, and even their British counterparts, were way out of my price range. In addition, building the guitar was a great challenge. My father had some experience with electronics, woodwork, and metalwork, so we very much enjoyed it, and believed we could build something better than what existed."

I followed up with a question I was extremely curious about: "Why did you become a musician after having completed your BSc in physics?"

May did not hesitate: "It was a call. I loved physics and astronomy, and the fact that I studied those subjects pleased my parents, but the call of music was so strong that I couldn't resist it. I was also afraid that if I didn't respond, it would never come back."

"Why, then, did you decide to return to your PhD studies in astrophysics after decades in music?" May reregistered for his degree after an interruption of thirty-three years!

"That was a very fortunate thing," May answered. "Even though I kept my interest in astronomy, it was really Sir Patrick Moore [a famous English amateur astronomer and popularizer of science], who was a 'father' to many astronomers of my generation, who asked me why I wouldn't go back. I didn't think that was possible, but I mentioned it in an interview, and suddenly I got a phone call from Michael Rowan-Robinson, the head of the astrophysics group at Imperial College. He told me that if I was serious, he would be my supervisor." May laughed again. "Being famous does open doors." He then continued, "This was not easy. You have to reenergize those parts of your brain you haven't used for a long time. Rowan-Robinson was very hard on me, which was very important, because the whole affair was very visible."

I thought to myself that energizing parts of the brain you don't routinely use is part of what curiosity is all about, and this naturally led to my next question: "Do you see any connection between your interests in music and in astrophysics? Or do they live in entirely separate worlds?"

May did not hesitate: "I think that my abilities in each field were definitely enhanced by my openness to the other field. I don't think that science and art need to be separate. In some mysterious way they are connected. For instance, I now know many scientists, such as Matt Taylor, the 'boss' of the Rosetta mission [a space probe launched by the European Space Agency to study Comet 67P/Churyumov-Gerasimenko], who are very interested in music."

"Why did you agree to become chancellor of Liverpool John Moores University?"

May laughed. "Because I was curious. I had no idea what would be involved, and I decided to check it out. I was also wondering if being a chancellor changes you. The answer, by the way, is no! It doesn't." He laughed again.

"And how did you become interested in Victorian stereophotography?"

"That is a passion I developed as a child and it has never left me. It's like magic."

"And your particular interest in 'Diableries' [a series of stereoscopic photographs supposedly depicting everyday life in hell]?"

"Those are incredibly labor-intensive works of art," May replied. "There is so much mystery and imagination in every work. Even with today's technology it is extremely difficult to reproduce any of these. I have just created with Claudia Manzoni a stereoscopic portrait of Comet 67P/Churyumov-Gerasimenko from images captured by the Rosetta mission, and also a 3D view of Pluto from images taken by New Horizons."

"Any other things you are curious or passionate about?" I asked.

May's response was swift: "Two things. First, animals: the cruelty that we have dished out to animals is horrible. I want to fight for their right to have a decent life and a decent death." He paused for a second, then continued. "The second thing I am infinitely curious about is human relations, and love in particular. Love is one of the most powerful things in our lives. It motivates us, and in ancient history entire empires were apparently built or destroyed because of it. Yet science has very little to say about love. It's the great fiction writers that come the closest to describing it."

I completely agreed with the last statement, but also thought that somewhat similar things could be said about curiosity. My last question was about an amusing anecdote that I had heard: "Astrophysicist Martin Rees once told you that he doesn't know any scientist who looks as much like Isaac Newton [especially because of the hair, and perhaps the nose] as you do. Did that ever occur to you?"

May laughed. "No. In fact, my first reaction when he said that was a bit of annoyance, because I was thinking, 'Is that all he wants to talk to me about?' But later we did have a wonderful conversation on astrophysics."

At the end, I asked May if he wanted to ask me anything.

He asked, "Are we alone?"

I reviewed for him the upcoming searches for extrasolar life. I also said that there is hope that within two to three decades we will either actually find some biosignatures—composition anomalies created by life—in the atmospheres of planets orbiting other stars, or will at least be able to place some meaningful probability constraints on the existence (or rarity) of extrasolar life. For me, the important point was that May was still genuinely curious about cutting-edge astronomical research.

Autodidacts

Their broad and diverse interests notwithstanding, each of the six people I had interviewed so far—Freeman Dyson, Noam Chomsky, Story Musgrave, Fabiola Gianotti, Martin Rees, and Brian May—is best known for his or her contributions in one particular area, the one in which they have formally studied or trained. Dyson is primarily known for his achievements in fundamental physics, Chomsky for his influential ideas in linguistics, Musgrave for being an astronaut, Gianotti for discovering the Higgs particle, Rees for his many contributions to astrophysics and cosmology, and May for being a virtuoso musician. My next interviewee is chiefly known for her brains.

Intelligence is a loaded word; it is difficult to define and even harder to measure. Nevertheless, from 1986 to 1989 Marilyn vos Savant was listed in the *Guinness Book of World Records* as having the "World's Highest IQ"—a staggering 228! While the precise numerical values of the scores in the Stanford-Binet and the Mega intelligence tests are notoriously unreliable, and vos Savant's specific scores have been called into question, no one has ever doubted her incredible intelligence. Amazingly, vos Savant never finished even an undergraduate degree, having studied philosophy for only two years at Washington University in St. Louis. Yet when *Parade Magazine* ran a profile of her, along with a selection of her answers to readers' questions, the response was so overwhelming that the magazine offered her a permanent job. In her weekly column, "Ask Marilyn," vos Savant answers a rich array of vocabulary and academic questions and presents and elucidates various puzzles in logic. Given her unusual background, I thought it would be interesting to compare vos Savant's perception of her own curiosity to that of some of the other interviewees. Consequently, I decided to concentrate on three main questions, starting with the one I was most curious about: "Over the years, which topics have you been most curious about? And why do you think those particular subjects aroused your curiosity?"

Being aware of the themes often covered in her column, I was expecting an answer related to probability theory or mathematical logic, but vos Savant surprised me: "I've long been curious about the human mind, the nature of consciousness, the breadth and depth of cognition, and the enigma of infinity. My cat doesn't know that she can't understand algebra. What don't we know that we can't understand, yet would be easily understood by an intellectually superior mind?"

I found this answer very interesting for two main reasons. First, unexpectedly, vos Savant was alluding to a slightly different variant of the famous "unknown unknowns" problem—those things that we don't know that we cannot understand. Second, the reference to an "intellectually superior mind" marginally touched upon another topic I am passionately curious about: whether there are any other intelligent civilizations in our Milky Way galaxy, and if they exist, what their nature might be. On one hand, since the age of the solar system (4.5 billion years) is less than half the age of the galaxy, another civilization, if it exists, could be more advanced than ours by more than a billion years. On the other hand, since there is still no compelling explanation for the "Fermi paradox"—the surprising lack of evidence for the existence of any such civilization—it could be that there are some evolutionary bottlenecks that make it extremely difficult to transit to intelligence.

My second question to vos Savant addressed the evolution of her personal curiosity: "Were you always curious? Have you noticed any changes in your curiosity over the years (during your adult life)?"

Her answer was very candid:

> When I was young, I was also curious about objects near and far—from frogs to the dwarf planet Pluto. That kind of curiosity has virtually disappeared, possibly because following those interests required a mentality that can aim through mi-

croscopes or telescopes, which I learned meant working with large (i.e., funded) scientific organizations. I can do the former, but my personality isn't suited for the latter!

Anyway, I'm much more interested in humanity now, especially in the way that many aspects of our lives are improving while, at the same time, great civilizations appear to be in various states of devolution. Fascinating! What does the future hold?

This reply was captivating, and perhaps representative of a common trend that is associated with the accumulation of life experiences. It appears that over the years, many people evolve from being interested in a variety of "objects" to being curious about all-embracing, more philosophical questions. Again, this may reflect a transition from a primarily perceptual or diversive curiosity to a state largely dominated by epistemic curiosity. As music critic and novelist Marcia Davenport once humorously wrote, "All the great poets died young. Fiction is the art of middle age. And essays are the art of old age."

My third question to vos Savant was the same one I had posed to a few of the other interviewees: "Have you noticed any other characteristic that is common to exceptionally curious individuals?"

Her answer was an interesting variation on the theme outlined by Gianotti: "I've noticed an ability to ignore the obvious—maybe because it isn't as interesting—and pay attention to seemingly insignificant aspects of subjects. Sometimes these less-evident facets are dead ends, but sometimes they explode in importance when poked by the right kind of individual."

Reflecting upon the combination of this insightful answer with that of Gianotti, I realized that it had *Feynman* written all over it. How else would you describe his fascination with phenomena that superficially looked mundane? I could also hear in vos Savant's remarks echoes of Dyson's declared interest "in the details" more than "in the big picture." Most important, however, vos Savant captured here an aspect of the essence of curiosity: not being interested in the *obvious*, preferring the obscure or the mysterious. As philosopher Martin Heidegger remarked, "Making itself intelligible is suicide for philosophy."

The next person I talked to, John "Jack" Horner, also never graduated from college. This fact has not stopped him from becoming one of the best-known paleontologists, a MacArthur Fellow, the science advisor to all of the *Jurassic Park* movies, and the discoverer of the charming fact that at least some species of dinosaurs cared for their young. He is also the person who showed that some dinosaurs previously believed to be different species simply represent the same dinosaurs at different ages.

I talked to Horner in September 2015. My first question was a bit tentative: "Would you call yourself curious?"

"Yes, that's *primarily* what I am," he replied straightaway. Horner discovered his first dinosaur bone when he was eight years old and excavated a dinosaur skeleton when he was thirteen. These remarkable unearthings naturally led to my second question: "How come?"

"My father was a sand and gravel person; he had a fairly good knowledge of geology. So he took me to a place where he thought I was likely to find dinosaur bones." After a short pause he added, "As it turned out, that became the first site where I made a few of my discoveries."

There was still something that wasn't quite clear to me. "Many children are fascinated by dinosaurs, yet most of them don't become paleontologists. How did you embark on the road to professional paleontology?"

Horner laughed. "I was extremely dyslexic. Even today, I can only read like a second grader. So, when the other kids were learning to read, I used to go out and search for fossils. Once I found something, I would go to the library, look at pictures of dinosaurs, and attempt to identify to which dinosaur those bones had once belonged."

I stopped him for a minute. "I assume that at that time no one really knew what dyslexia was?"

"Indeed," he replied. "Some people thought I was retarded. My father believed for the longest time that I was just lazy. In fact," he laughed, "he continued to believe that until my photograph appeared on the cover of his favorite magazine."

I told Horner that this amusing story concerning him and his father reminded me of an interview I once saw on TV with the father of Barry, Robin, and Maurice Gibb, the brothers who formed the pop music group the Bee Gees. The interview took place at just about the time the Bee Gees, who wrote all of their hits, were at the peak of their success. Yet the father insisted, "These boys never worked one day in their lives."

Since I knew that Horner did attend some classes in geology and zoology at the University of Montana, I asked him to describe that experience.

"I went to the university for several years and learned a lot, but I could never pass the tests, because essentially all exams required extensive reading," he recalled.

"What, then, did you actually learn?" As soon as I asked the question, I realized that I could have anticipated the answer.

"The university had a nice collection of fossils and I was very curious about those."

"Still," I wondered, "in research today, it is difficult to make progress if you cannot read, isn't it?"

Horner laughed loudly. "I always tell my students: 'If you do it first, you don't have to read anything.'"

In addition to being funny, this answer left me breathless. Without knowing it, Horner was almost exactly quoting Leonardo. Recall Leonardo's reaction to the accusation that he wasn't well read: "Those who study the ancients and not the works of Nature are stepsons and not sons of Nature, the mother of all good authors." Just as Horner did five centuries later, Leonardo exclaimed, "Though I can not like others cite authors, I shall cite much greater and more worthy: experience, the mistress of their masters."

Horner continued to reiterate the same sentiments: "What I discovered in my research was that many other scientists had preconceived ideas based on what they had read. I didn't have any. When I discovered something, I wrote about what I found and what conclusions I personally drew from those findings." Horner was obliquely touching here on another somewhat unfortunate reality that vos Savant also alluded to. Few scientists today can afford to take risks and independently follow their curiosity because the competition for funding and recognition is fierce. The more expensive the science is, the more it may discourage individual curiosity and "outside the box" exploration in favor of incremental progress.

Coming back to the question I had asked a few of the other "curious minds," I inquired, "Can you think of any other characteristic that goes hand in hand with curiosity?"

"Excellent question," he replied. "Maybe you can identify what it is, if I'll tell you that right now I am putting together a talk that will be presented in the context of a course entitled Introduction to Biotechnology. Confidentially, let me tell you," he whispered dramatically, "that I find many of the other talks in this course to be somewhat on the dry side. My topic is [here his voice rose again] 'how to make a glow-in-the-dark pink unicorn.'"

To make sure I correctly understood the topic, I inquired with some disbelief, "You are seriously talking about making a new live species—a live pink unicorn that glows in the dark?"

"Indeed. Some people are driven to success—they may have the desire to cure cancer. I am curious about this theoretical problem: Can we actually make one? How many things do we have to know in order to make one?"

This was nothing short of a mind-blowing exercise, which also fit perfectly with Gianotti's concept of "the ability to think beyond" and vos Savant's notion of "the ability to ignore the obvious." "And this sums up your philosophy of what curiosity and science are all about?" I asked.

Horner again was very confident. "I think that the best kind of science emerges when you follow your personal curiosity rather than following anybody else. Your only goal should be to try to satisfy your curiosity."

I happened to know that Horner was involved in yet another big project, so I felt that I should ask him about that one too: "How about the Reconstructing a Dinosaur project?"

Horner had expected that question. "Unlike other attempts, we are not using ancient DNA." He was referring to fascinating work by Harvard geneticist and molecular engineer George Church, aimed at "de-extincting" the woolly mammoth by using genetic segments from frozen mammoth specimens. "Rather," Horner continued, "we use bird DNA and try to retro-engineer it. As it turns out, making a tail captures much of the difficulty, because it essentially involves making vertebrae."

Astonished by the ambition of this endeavor, I could only remark, "It would be pretty amazing even if you only partially succeed." Given the intellectual chutzpah of Horner's projects, I couldn't refrain from asking a last question: "Do you choose your graduate students and postdoctoral fellows from among those who share your level of curiosity?"

"Absolutely!"

My last interviewee might not have become a world-famous artist were it not for the fact that he was shot in the leg. Here is how Brazilian sculptor, photographer, and mixed-media artist Vik Muniz describes the events of that fateful night in São Paulo:

> One night, after leaving a social event, I witnessed a fight between two men, one of whom was beating the other violently with brass knuckles. I got out of my car and helped separate the victim from his assailant, who ran away. Walking back to my car, I heard a great explosion and suddenly I was on the floor, crawling for my life. The victim, incapable of clear judgment, had opened his car door, reached for his gun, and unloaded the entire barrel in the direction of the first person he saw wearing dark clothes. That person was me. Luckily, the shot wasn't fatal. And even more luckily, the gunman was a rich man. Begging me not to press charges, he offered me a reasonable sum of money. I used it to buy a plane ticket to Chicago in 1983.

Muniz is currently based in New York, although he spends time in Rio de Janeiro. He is an artist with a pyrotechnic imagination, best known for meticulously and shrewdly re-creating iconic artworks with everyday materials, such as chocolate syrup, sugar, diamonds, and peanut butter, and then photographing them to produce photojournalistic-style images.

In 2010, the film *Waste Land* documented an ambitious project Muniz undertook at the world's largest landfill, Jardim Gramacho, on the outskirts of Rio de Janeiro. In that endeavor, he collaborated with waste pickers (known as *catadores*) to literally transform garbage into art. *Waste Land* was nominated for an Academy Award and won more than fifty international prizes.

When I spoke with Muniz in February 2016, I asked him about something I learned from his book *Reflex*: "I know that you like Ovid's narrative poem *Metamorphoses.* Can that be taken as the motto of your entire work?"

Muniz laughed. "Maybe not quite a motto, but an inspiration. You know, the first phrase in *Metamorphoses*, 'My mind is bent on telling stories of bodies changed into new forms,' is such an interesting statement about perception and interpretation." After a brief pause, he continued, "Both artists and scientists try to look at everything with wonder. For years I tried to find a definition for art, and I finally came up with 'a development or evolution of the interface between mind and matter.'" He laughed again and said, "Then I realized that the same definition could apply to science, too."

"Do you see more connections between the arts and science?" I asked.

"Definitely," Muniz responded right away. "Both scientists and artists are 'hungry'—they dedicate their lives to the creative tools that will help us find out what is out there. When I talk to scientists, I am impressed by the fact that, for example, in the subatomic world, they think about things that are beyond the realm of the senses. How do you perceive or understand dimensions beyond the three dimensions of space? This is hard for people who are used to thinking visually."

Muniz's remark was very similar to Gianotti's description of curious individuals as people "with the ability to think beyond" and to May's comment that science and art are connected "in some mysterious way." This naturally led to my next question: "Do you see yourself as a curious person?"

Muniz laughed loudly. "You could say that my curiosity is so profound that it's almost like a sickness. When I was a child, someone gave me a screwdriver as a gift, and I almost dismantled the entire house. They had to take it away from me, because I even got electrically shocked. I don't consider myself erudite, but I do try to know at least a little about almost everything. I feel that the seed of creativity is curiosity, and that the potential for imagination comes from wondering." He was silent for a few seconds and then added, "Sometimes I almost envy the people of medieval times, when so little was known and there was a whole world to be curious about."

"I want to ask you about two of the things that I know fascinate you: light and comedian Buster Keaton."

Muniz explained, "In my work, I try to uncover how we can translate the information we get from the senses into a mental picture. There is so much they don't teach you in art school: the physics of light, the physiology of vision, the neuroscience and psychology of vision. You cannot work without knowing all that. Consequently, half of my library in New York is composed of science books."

This was precisely Leonardo's attitude. "And what about Buster Keaton?" I continued.

"Two main things characterize his work: mechanics and cause and effect. It's the mechanics of humor and the mechanics of the body, which in silent films was so much more important. I simply think Keaton is brilliant."

I knew that one of the artworks in his series *Pictures of Ink* is a portrait of Richard Feynman (Figure 23). In this series Muniz created hand-made renderings in thick ink of well-known images. "Why Feynman?" I asked.

"I read all of his popular books," Muniz told me. "Every scientist I know was deeply impressed by Feynman."

Indeed, I thought.

"He even came to Brazil to learn how to drum," Muniz continued. "He had that mode of observation that was very open. Both

Figure 23

scientists and artists have to have that, to be able to invent new ways of looking at things."

All I could think of was *indeed* again. Finally, I asked Muniz what inspired him to do the project at the Jardim Gramacho landfill. I found his answer sincere and quite moving.

"For me it was one of those moments. I was working on a career retrospective, and I said to myself, 'I know what art did for me,' but I was wondering what it did for other people. So I started working with people who had no real previous connection to art. Ultimately, it was largely driven by curiosity." The money that was raised from auctioning the resulting artworks was given to the Brazilian *catadores*.

A Vigorous Mind

Samuel Johnson wrote in 1751, "Curiosity is one of the permanent and certain characteristics of a vigorous mind." If we examine the responses of the prodigiously curious individuals I have interviewed, can we gain any insights from their personal stories and their undoubtedly vigorous minds? I suspect we can.

While childhood memories should always be taken with a grain of salt since they may be subjected to later corrections and embellishments, the accounts I have gathered leave little doubt that even if they have never consciously thought about it, people who are exceptionally curious as adults were also unusually curious children. Not every child attempts to solve the mystery of the tides (as Martin Rees did), and while many children play with dinosaur toys, very few actually dig for dinosaur bones (as Jack Horner did). Let's hope that fewer and fewer get electric shocks while being curious, which was Vik Muniz's experience. Curiosity manifests itself in the form of keen interests and enthusiasm for exploring phenomena, events, or artifacts. But it is also fairly clear that having insatiable curiosity does not necessarily mean the child will be identified as "gifted" (consider Horner's story).

Psychologist Mihaly Csikszentmihalyi speculated that children become interested in pursuing those activities that give them an edge in the competition for the attention and admiration of the adults that are significant in their lives. Thus he argued that a girl recognized for her ability to jump and tumble is likely to become interested in gymnastics. While this scenario certainly applies in some cases, such as Picasso's, who showed an incredible talent in drawing at an extremely young age, the picture can be much more complex (as in the case of Fabiola Gianotti or Marilyn vos Savant, for instance). Brian May's path zigzagged: he participated in building a guitar with his father, then studied math and science, left that route (in spite of his parents' objection) for music, only to eventually return to science. There is another important lesson here: individuals can keep their curiosity alive for many years, and even get back to topics that interested them in an earlier stage in life. Csikszentimihalyi himself acknowledged that more often than not, the competitive edge itself is not the result of heredity. Instead, the early curiosity may be sparked by very particular circumstances in the child's immediate environment.

The accounts of Gianotti and Rees of their college years show that not every curious or even highly accomplished scientist committed to a scientific career from the very start. Rather, as Jacqueline Gottlieb's experiments indicate, some explored a broader intellectual panorama before settling into and concentrating on one particular passion. A most extreme example of shifting interests and wandering curiosity was provided by the incredible trajectory of Story Musgrave. His course, with one curiosity inspiring another, was remarkably similar to that of chemist and Nobel laureate Ilya Prigogine. Despite the fact that his original main interests were in the humanities, under pressure from his family Prigogine started to study law. This led to interest in the psychology of the criminal mind, followed by his delving into neurochemistry in an attempt to decipher the underlying brain processes. Realizing that neuroscience was still far from being able to fully explain behavior, he decided to start with the foundations and immersed himself in the basic chemistry of self-organizing systems.

Recall that Musgrave also went from mathematics to computer science to chemistry to medical school, eventually becoming a celebrated astronaut. The implication is that, on one hand, curiosity provides the guiding light, but on the other, it can illuminate a winding road. Outstandingly curious individuals may not be able to predict where curiosity will lead them (as in the cases of Dyson, vos Savant, Muniz, and May), but at all times they remain attentive to the world around them and prepared to attempt to solve some of its mysteries. One characteristic that seems to keep curiosity fresh (at any age) is a certain openness to recognizing unfamiliar problems even in new domains. Rees's interest in existential threats, May's passionate activism for animals, and Horner's investigation into how to make a pink unicorn are excellent examples. Maybe even more impressive, in an interview with *Quanta Magazine* a few days after his ninetieth birthday, Freeman Dyson revealed that he had taken on a new challenge: to formulate a mathematical model for effective clinical trials with minimal loss of life. How is that for maintaining and utilizing one's intellectual energy?

NINE

Why Curiosity?

HUMAN CURIOSITY HAS CLEARLY EVOLVED AT least partially to aid survival. An understanding of the world around us, its causal connections, and the sources of changes have helped humans to reduce prediction errors, cope with the environment, and adapt. Curiosity about other humans has no doubt played a role in mating and in the creation of social structures. The eighteenth-century adventurer Giacomo Casanova is often quoted as having said: "Love is three-quarters curiosity." In fact, what he said in his *Memoirs* is: "The woman who showing little succeeds in making a man want to see more, has accomplished three-fourths of the task of making him fall in love with her; for is love anything else than a kind of curiosity?" At the same time, the craving for knowledge for its own sake and curiosity about a whole trove of abstract concepts have led to a rich and sophisticated human culture.

Humans don't just passively react to what they see, hear, or feel. They show interest in phenomena near and far, and occasionally they actively engage in exploration. In a relatively small number of people, certain topics stimulate such a powerful epistemic hunger that they devote their entire lives to the pursuit of answers. However, people are not all equally curious. Undoubtedly, the level of curiosity an individual expresses is to some extent, if not mostly, genetically dictated. In fact, there exists considerable experimental evidence suggesting that essentially all the psychological traits are heritable. Still, it is interesting to attempt to understand to what degree other factors play a role in determining how curious people are. What is ultimately responsible for those non-innate "individual differences," and even collective trends? Factors other than genetics could include, for instance, the influences of one's immediate family, of close friends, of teachers, of religious institutions, and of the general cultural environment and heritage. Understandably, it is not always easy to separate the genetic and environmental effects, especially since the two sometimes intricately interact. For example, while it is certainly true that a chain of tragic events in someone's life may sink him or her into deep depression, it is also well established that the genetic makeup of some individuals makes them more susceptible to depression than others, even under very similar circumstances.

Heritability and Curiosity

In order to obtain a cleaner estimate of the heritability of various psychological characteristics, including curiosity, researchers such as Thomas Bouchard at the University of Minnesota and Robert Plomin and Kathryn Asbury at King's College, London, rely largely on studies with twins. Typically, one-third of all twins are identical (and therefore genetically equivalent), and the rest are divided equally between those of the same gender and of different genders. Bouchard and his colleagues are best known for an influential research project known as the Minnesota Study of Twins Reared Apart (MISTRA), which brought together twins from all over the world who had been separated during childhood and for most of their lives until that point. Plomin leads the Twins Early Development Study, an effort that involves about 12,000 families, on which Asbury had also worked.

The MISTRA twins were subjected to about fifty hours of psychological and medical examinations, with special emphasis on mental ability including tests such as the Wechsler Adult Intelligence Scales and Raven's Progressive Matrices. The results were quite conclusive: identical twins who spent much of their lives apart were essentially as similar in intelligence as those who grew up together.

In 2004, Bouchard reviewed the results of a number of large projects performed with broad samples drawn from relatively affluent Western societies. The findings revealed that genetic influence was in the range of 40 to 50 percent for all the Big Five personality traits (openness, conscientiousness, extroversion, agreeableness, neuroticism), with openness (the characteristic most related to curiosity) scoring as high as 57 percent in heritability. In other words, genetics could explain about half of the observed differences in personality traits. No significant differences in heritability were found between the two sexes.

Bouchard also examined data gathered over many years in another large study that concentrated specifically on psychological interests (also called occupational interests). That particular research project involved twins, nontwin siblings, and parents and their children. It inquired about interests expressed in the artistic, investigative, social, and enterprising domains. Of these, an investigative interest is clearly indicative of curiosity, although all the other interests probably also involve an important curiosity component. Again, all of these inclinations *showed significant genetic influence, at an average level of 36 percent*, with a modest shared environmental influence at about 10 percent for each one of the traits.

Is the strong genetic influence on curiosity surprising? Probably not. As we have seen in chapters 4–6, curiosity requires certain cognitive abilities, and it may depend on working memory capacity and executive control, all of which are governed to a significant degree by genetic inheritance. Here again, however, without the proper exposure to opportunities and the availability of psychic energy that is not fully committed to survival and the necessities of life, the genetic characteristics may remain latent. Bouchard himself noted, in this respect, that "because the studies probably undersampled people who live in the most deprived segment of Western societies, the findings should not be considered as generalizable to such populations." Even more important, we know that genetics cannot tell the whole story. A world following only the instructions encoded in our genes by evolution would be very different from our world. It would probably have no Shakespeare, no Mozart, and no Einstein. Dramatic developments, such as the appearance of human language, the historical circumstances that led to the Renaissance, and the scientific revolution, all brought about at least partly by human curiosity, allowed humans to take a speedier route than the one paved by DNA alone. What we refer to as our "culture" was born out of taking advantage of this non–biologically bound curiosity highway. Instead of evolving solely through mutations in human genes (a process that is painfully slow), human civilizations have evolved through the acquisition and dissemination of knowledge. There was still an important selection process of useful information that the human mind had to perform, and that's where some of those curiosity and exploration strategies that I discussed in chapter 5 came into the picture. The environment bombards our senses with data, from which our brains have to continuously choose those pieces needed for survival and for the satisfaction of our specific, diversive, perceptual, and epistemic hungers.

Given the important role that curiosity plays in such diverse areas as education, basic research, artistic aspiration, and storytelling in all of its different guises (interpersonal communication, books, films, advertising, etc.), even if we accept as a fact the notion that a significant part of the individual differences in curiosity has a genetic origin, the question still arises: Can one actually cultivate curiosity? Before we examine potential ways to enhance curiosity, however, we must recognize the reality that there are circumstances that can act to strongly suppress it.

Curiosity Killed the Cat

People who have to struggle to survive don't have the luxury, the motivation, or the time to contemplate the meaning of life. The small children of refugees, forced to cross borders and sometimes entire continents on foot while experiencing unrelieved hunger and lack of proper shelter, can hardly be expected to engage in any exploration or activity that provides reward only for its own sake.

In addition, there were entire periods in human history during which the myths, traditions, and sometimes deliberate misinformation labeling curiosity as perilous, served as strong deterrents. Oppressive rulers, harsh imposers of strict religious orthodoxy, controllers of information, and in general staunch guardians of the status quo sometimes felt that their subjects should be inferior to them in knowledge, and therefore that curiosity should not be encouraged. Convincing the masses that what you don't know won't kill you and that things are the way they are because that's how they should be apparently was easier for some of these individuals in power than to actually acquire superior knowledge through learning.

There has probably never been a civilization that did not build walls around some types of knowledge. The tradition that curiosity can be dangerous and therefore it shouldn't be allowed free rein is as old as human culture itself. In the Bible, Eve and Adam are banished from the Garden of Eden for yielding to their curiosity (incited by the crafty serpent), wanting to know more than they should, and eating the forbidden fruit. The Scottish playwright known by the pseudonym James Bridie humorously (or seriously?) described Eve's actions as "the first great step in experimental science."

Also in the book of Genesis, when God decided to destroy the sinful cities of Sodom and Gomorrah, he nevertheless resolved to spare the lives of the virtuous Lot, his wife, and their two daughters. Two angels were therefore dispatched to urge Lot to immediately leave the city of Sodom and to not look back under any circumstances. Lot's wife succumbed to her curiosity and glanced back, only to be instantly turned into a pillar of salt. (Just as an aside, she had to be a very large person to correspond to the dimensions of the rock formation in Israel traditionally known as "Lot's Wife.")

The notion that some knowledge is illegitimate and forbidden to all humans continued to pepper other texts in the Scriptures and in a variety of theological manuscripts. In the canonical Wisdom Book of Ecclesiastes, for instance, we find the discouraging warnings "In much wisdom there is much grief, and increasing knowledge results in increasing pain," as well as the admonition "Be not curious in unnecessary matters: for more things are shewed unto thee than men understand." You can hear a later echo of this deterrent in St. Augustine's fifth-century proclamation "God fashioned hell for the inquisitive." St. Augustine also referred to curiosity as "lust of the eyes" (*concupiscentia oculorum* in Latin), and he warned against attempts to count the stars or grains of sand, since such vain curiosity, he asserted, created an obstacle on the path to humble devotion. These sentiments strongly resonated with the twelfth-century French abbot St. Bernard of Clairvaux, who elevated curiosity to the status of a deadly sin, situated somewhere between sloth and pride. "To learn in order to know is scandalous curiosity," he pronounced.

Curiosity didn't always meet with the approval of the ancient Greeks, either. Greek mythology contains a number of stories of divine punishments inflicted upon those who were too curious. In a legend strikingly similar to the account concerning the biblical Eve, Pandora was not able to resist her curiosity and opened a jar (commonly mistranslated as a box), thereby releasing all the evils of humanity. Severe punishment fell on the princess sisters Herse and Aglauros, who, overcome with curiosity, disobeyed Athena's specific orders and peeked inside the intriguing basket that contained the infant Erichthonius. The sight of this mythical future ruler of Athens (who, according to some versions, was half-human and half-snake) drove the sisters insane, and they threw themselves off the Acropolis. The myth of Semele, who was curious enough to insist on seeing Zeus in all of his divine glory, in spite of his imploring her not to make this request, ended also in disaster: she was consumed by a lightning-ignited fire.

You will notice, though, that in most of these cases, one could argue that the act that invoked the penalty was really disobedience rather than curiosity. We should also remember that until about the seventeenth century, the meaning of curiosity was somewhat different from the one we have today. For humans to be curious implied to various classes of presumed moralists that they were prying into affairs that were none of their business rather than that they were exploring. Accordingly, the twelfth-century English scholar Alexander Neckam mocked even human inventions and achievements in architecture as acts of meddling in God's creations: "O vain curiosity! O curious vanity! Man suffering from the illness of inconstancy, 'destroys, builds, and changes the square to round.'" Even the great Dutch Renaissance humanist Erasmus of Rotterdam, who generally insisted that "the words [in the Scriptures] do not condemn learning," contended that curiosity represented a greed to know unnecessary things and that therefore it must be the province of the elite.

The general attitude toward curiosity started to change in the sixteenth century, especially with the increase in the number of world travelers and naturalists. In fact, questions such as who ought to know what, and how they were supposed to acquire that knowledge, became the subjects of conversations in circles ranging from scientific to religious societies. Oxford historian Neil Kenny discovered that even such a simple gauge as the number of times words like *curiosity* and related words derived from the Latin *curiositas* were used in a broad range of literature soared about tenfold between 1600 and 1700. This reflected the growth in the interest in exploration sparked by the scientific (and indeed also philosophical) revolution. The first person to recognize curiosity as an emotion from which humans cannot escape was the restlessly curious French mathematician and philosopher René Descartes. While his inclination to think of curiosity as akin to sickness demonstrates that even he was still ambivalent toward this passion, he nevertheless pronounced, "So blind is the curiosity by which mortals are possessed that they often conduct their minds along unexplored routes, having no reason to hope for success, but merely being willing to risk the experiment of finding whether the truth they seek lies there." When he created his inventory of six "primitive passions," Descartes listed *wonder* (which is intimately related to curiosity) first. He explained that the function of wonder was to "learn and retain in our memory things of which we were previously ignorant."

Other notoriously curious characters followed. The idiosyncratic English physician and writer Thomas Browne, for instance, published books on topics as diverse and esoteric as the mysteries of nature, humans and their relation to God, beliefs and superstitions, antiquarian subjects, the history of horticulture, and death.

At the beginning of the nineteenth century, the Prussian naturalist and explorer Alexander von Humboldt traveled extensively in

South America, Russia, and Siberia and published detailed works in botany, anthropology, meteorology, geography, archaeology, and linguistics. One of his biographers wrote that Humboldt "took the world as a laboratory to explore." Humboldt's brother Wilhelm, a linguist and philosopher himself, remarked that Alexander had "a horror of the single fact," preferring instead to explore every aspect of a phenomenon. It would probably not be an exaggeration to say that Humboldt personified curiosity itself. In the introduction to his multivolume work *Cosmos*, in which he attempted to outline all the available knowledge about the physical sciences, Humboldt emphasized the egalitarian nature of curiosity by writing that scientific knowledge was "the common property of all classes of society." Using almost verbatim the words that Leonardo recorded 300 years before him and Feynman 150 years after him, Humboldt expressed what could almost be taken as a curious person's manifesto: "There is nothing that does not interest a naturalist as long as he makes a detailed study. Nature is an inexhaustible source of study, and as science advances so new facts reveal themselves to an observer who knows how to interrogate her." Late in life, Humboldt referred to his own unquenchable curiosity: "I like to think that, while I was at fault to tackle from intellectual curiosity too great a variety of scientific interests, I have left on my route some trace of my passing." Oxford social historian Theodore Zeldin beautifully summarizes Humboldt's contribution: "He dared to make a link between knowledge and feeling, between what people believe and do in public and what obsesses them in private."

In spite of the more positive light in which curiosity was seen from the seventeenth century on, many people remained wary of it. An excellent example of this mistrust is Goethe's nineteenth-century tragic play *Faust*, in which a German scholar sells his soul to the devil after being frustrated by his own endeavors to acquire knowledge. The same period also witnessed a phase in which the word *curiosity* came to characterize not just the human thirst for information but also those rare or exotic objects that people became interested in. This led to the emergence of "cabinets of curiosity" (or "wonder rooms"), effectively small museum-like collections of items from the natural world or the arts.

Equally revealing is the fact that in their collection of fairy tales published in 1812, the Brothers Grimm included many stories with an ambiguous message about curiosity and the drive for exploratory behavior. In their variant of *Sleeping Beauty* (which was based on a tale originally published in 1697), the fifteen-year-old princess eagerly investigates all corners of her castle, finally arriving at a small tower. After climbing the winding staircase and opening a little door with a rusty key, she finds herself in front of an old woman spinning her flax. The wide-eyed princess barely touches the spinning wheel when the spindle pricks her finger, causing her to sink into a profound slumber for a hundred years. Hardly an encouragement to conduct curious explorations!

In the tale of *Hänsel and Gretel*, the young brother and sister find themselves in a similarly dramatic predicament, when their adventurous journey leads them to a house constructed of cake and confectionery. Not knowing that the house belongs to a cannibalistic witch, the two children endanger their lives by starting to nibble at the house's roof. The witch, by the way, is reminiscent of the long-nosed supernatural being Baba Yaga of Slavic folklore, who also eats nosy children.

Even though both *Sleeping Beauty* and *Hänsel and Gretel* have happy endings (the princess eventually gets her prince; Hänsel and Gretel save their lives by outwitting the witch), these fairy tales, and many others, seem to imply that curiosity is hazardous. That is also the message encapsulated in the common proverb "Curiosity killed the cat." Interestingly, the original version, which first appeared in print at the end of the sixteenth century, was "Care killed the cat," where the word *care* referred to grief or worry. It is not clear (at least to this author) how *care* got replaced by *curiosity* around the end of the nineteenth century, but the cautionary expression obviously meant to serve as a warning against inquiry and as advice that it's best to mind one's own business.

Since curiosity is not only inevitable but is also a principal driver of the desire for the acquisition of knowledge, we can perhaps take some comfort in the fact that one version of the idiom "Curiosity killed the cat" has the more positive rejoinder "but satisfaction brought it back."

Curiosity Is the Best Remedy for Fear

Unfortunately, the impediments to curiosity are not exclusive to biblical or medieval times or to ancient Greece. Tyrannical, iron-fisted regimes and ideologies and narrow-minded societies still attempt forcibly to put an end to curiosity even today.

Acts aimed at smothering inquisitiveness, novel ideas, and exploration are not limited to discouraging the sciences. The arts, and knowledge in general, have not been spared. In 1937, for instance, the Nazi regime organized in Munich the *Degenerate Art* exhibition, whose sole purpose was to convince viewers that modern art represented no less than a malicious plot by Jewish Communists against the German people. The exhibition included works by some of the greatest artists of the twentieth century: surrealists such as Max Ernst and Paul Klee; expressionists such as Ernst Ludwig Kirchner, Emil Nolde, Oskar Kokoschka, and Max Beckmann; cubist-symbolists such as Marc Chagall; abstract painters such as Wassily Kandinsky and Ernst Wilhelm Nay; and many others. Paintings were deliberately slapped on the walls in no logical order to convey the impression of worthlessness. In the exhibition catalog, abstract paintings were introduced by derogatory descriptions such as "There is no telling what was in the sick brains of those who wielded the brush or the pencil." To intensify the negative public reactions, the organizers hired agitators who mingled with the visitors and loudly mocked the art. Some of the works were even burned later.

This was not by any means the last time a reactionary, intolerant, or totalitarian regime destroyed art or took deliberate steps to stifle curiosity. On March 14, 2001, the theocratic Taliban government in Afghanistan announced the dynamiting and obliteration of the two great Buddhas of Bamiyan. These monumental statues (about 175 and 125 feet tall; Figure 24 shows the smaller Buddha

Figure 24

in 1977), were constructed around the sixth century. At the time of the destruction, the Taliban also shattered statues in the Kabul Museum and other museums in the Afghan provinces, thus annihilating historical links with Afghanistan's past.

The most appalling Taliban attack on curiosity, however, targeted a curious person: Malala Yousafzai. Born in 1997 in Mingora, Pakistan, Malala became a known activist during her childhood years. In 2008, following attacks by the Taliban on girls' schools, she gave a talk entitled "How Dare the Taliban Take Away My Basic Right to Education?" This courageous act was followed by her blogging for the BBC. The Taliban issued a death threat against Malala when she was fourteen, and on October 9, 2012, a gunman shot her in the head as she was riding the bus home from school. Fortunately she survived, went on to win the Nobel Peace Prize in 2014, and continues her vocal advocacy for the education of girls. In July 2015, this young, courageous, and curious activist opened a school for Syrian refugee girls in Lebanon.

A classic form of extreme censorship and curiosity suppression is book burning. Accounts of various biblioclasms date back to the seventh century BCE, but book-burning events continued well into the twentieth century. The Nazis, for instance, regularly engaged in the incineration of books by Jewish authors. The Chilean fascist dictator Augusto Pinochet ordered the burning of hundreds of books in 1973. In 1981, as part of a three-day pogrom against the minority Tamil population, Sinhalese police and government-sponsored paramilitaries burned down the Jaffna Public Library in Sri Lanka, which contained tens of thousands of Tamil books and manuscripts.

Is there a lesson in these stories of oppression, intimidation, and assaults on personal freedom? I strongly believe there is, and it is a fairly palpable one: *Curiosity is the best remedy for fear.* One of the clearest manifestations of freedom is precisely the ability to become interested in anything you like. Freeman Dyson noted this fact in the narrower sense of its application to science when he said, "Being a scientist gives you the 'license' to work on any scientific problem." However, freedom really means that you can follow your curiosity to wherever it takes you, as long as you don't infringe on other people's freedom and you're guided by certain ethics (I discuss this topic further in the epilogue). Or, as Oxford scholar Theodore Zeldin astutely puts it, "Being interested in one's work, in a few hobbies, in a few people, leaves too many black holes in the universe."

I coined the phrase "Curiosity is the best remedy for fear" when preparing a public talk in 2012. Shortly thereafter, however, I discovered that I was not the first person to have thought of that "therapeutic" property of curiosity. The tag line for the 2008 exhibition *U-Turn Quadrennial for Contemporary Art* in Copenhagen was the very similar phrase: "Replace Fear of the Unknown with Curiosity" (Figure 25). This expression ultimately means that, in the same way

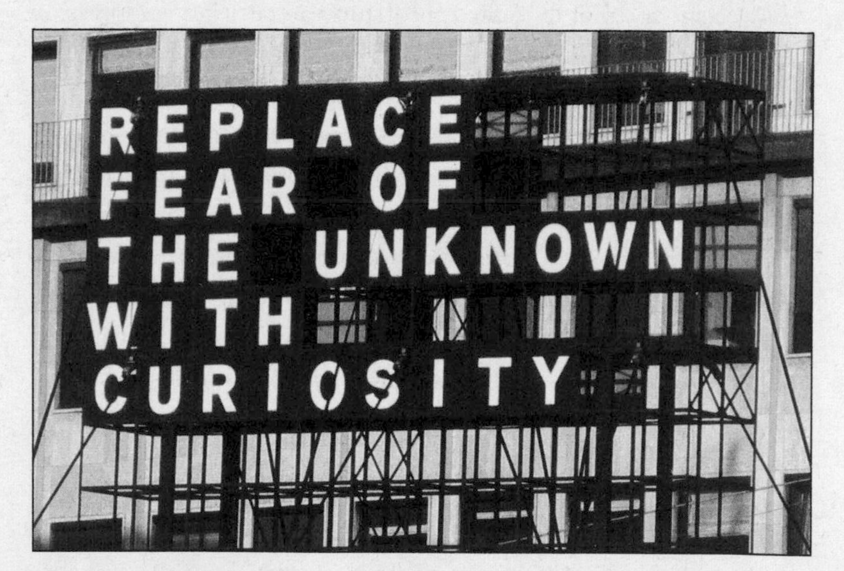

Figure 25

scientists have continued to discover ever since the scientific revolution that every breakthrough introduces a new series of questions and uncertainties, we should realize that the world around us offers an infinite number of opportunities to be curious and an abundance of topics to be curious about. We shouldn't allow our curiosity to be muzzled. In the words of Vladimir Nabokov, "Discussing these matters implies curiosity, and curiosity in its turn is insubordination in its purest form."

During the process of writing this book I unexpectedly stumbled upon the fact that the Irish novelist James Stephens used an even more powerful phrase than "the best remedy for fear" to describe curiosity's strength. In a philosophical novel entitled *The Crock of Gold*, he describes a boy who grows up in a thick forest in which the sunlight never penetrates. Not far from his home, though, the boy discovers a clearing where for a few hours in the summer the sun blazes down. "The first sight of the extraordinary blaze astonished him," Stephens writes, and then, as if echoing Leonardo at the entrance to the cave, he continues, "He had never seen anything like it before, and the steady, unwinking glare aroused his fear and curiosity equally." Stephens concludes with this forceful line: "Curiosity will conquer fear even more than bravery will; indeed, it has led many people into dangers which mere physical courage would shudder away from, for hunger and love and curiosity are the great impelling forces of life."

As it turns out, the intricate interrelation between curiosity and fear is more than the subject of a motivational statement. It has a physiological basis. The neurotransmitter dopamine has been implicated in both reward (and thereby curiosity) and fear in adjacent regions of the brain. In 2011, University of Michigan psychologists Jocelyn Richard and Kent Berridge showed that when dopamine was allowed to act normally, injection of the chemical into the front part of the nucleus accumbens of rats caused them to eat nearly three times more than usual. In contrast, when dopamine was injected at the back of the nucleus accumbens, the rats reacted fearfully, as if chased by predators. These experiments demonstrate that not only figuratively but at some level literally curiosity can cross the thin line between dread and reward.

Having examined depressing historical examples of the collective suppression of curiosity, we can now return to a more uplifting and fascinating question: How can we stimulate and cultivate individual curiosity, elevate it, and keep it vibrant? I should emphasize that I do not intend this next section to be a comprehensive "how-to" or "self-help" text. Rather, I shall distill a few of the lessons from the previous chapters into ideas that can assist our innate sense of curiosity.

Fanning the Burning Desire to Know

In his entertaining book *What Do You Care What Other People Think?* Richard Feynman tells a charming story of how during his childhood his father did his best to provide him with the mental tools that eventually helped turn Feynman into a scientist with an extraordinarily inquisitive mind. The story itself appears superficially to be unsophisticated. His father drew Feynman's attention to the fact that a certain bird was walking around pecking its feathers all the time (he probably meant "preening," rather than "pecking"), and he asked the boy why he thought birds did that. Feynman answered, "Well, maybe they mess up their feathers when they fly, so they're pecking them in order to straighten them out." The father suggested a simple way to test this hypothesis. He pointed out that if Feynman's conjecture was correct, one would expect that birds that had just landed would peck (preen) their feathers much more than birds that had been walking on the ground for a while. The father and son watched a few birds and concluded that there was no discernible difference between birds that had just been flying and those that had not. Feynman acknowledged that his hypothesis was probably incorrect, and he asked his father for the correct answer. His father explained that birds were bothered by lice that eat a protein that comes off the feathers. There are mites that eat some waxy stuff on the lice's legs, and in turn some bacteria grow in the sugar-like material that the mites excrete. He concluded, "So you see, everywhere there's a source of food, there's *some* form of life that finds it."

This seemingly innocent story of a childhood memory is remarkable in a number of ways. First, his father taught Feynman the pleasure of observation and wonder. As Feynman himself put it, "I'm always looking, like a child, for the wonders I know I'm going to find—maybe not every time, but every once in a while." Second, the father generated specific-perceptual curiosity by pointing out an intriguing phenomenon—birds preening their feathers—and asking a question about it. His father created in Feynman's mind an information gap that appeared to be surmountable—a sure way to pique curiosity. Similarly, a child who can name forty-two of the U.S. states is more likely to be interested in learning the ones she misses than a child who barely knows five. Third, his father did not give the answer immediately; rather, he encouraged epistemic curiosity by putting forward a test of Feynman's proposed explanation. Moreover, recall that experiments have demonstrated that when your theories turn out to be wrong, you are more likely to remember the correct interpretation (and even your incidental memory is enhanced). Finally, his father gave an answer Feynman knew even then to be probably incorrect in its details—birds preen to remove dust and parasites, to align feathers in an optimal position, and to distribute oil that is secreted from a gland—but that was right in principle. His father also used this common example of birds preening to convey a glimpse of the much bigger picture of life, its processes, and its dependence on food resources in nature, again promoting the development of epistemic curiosity.

Feynman's story contains, therefore, a number of important general clues as to what one can do to nourish curiosity, both internally and in others. First, it is important to endeavor to preserve the ability to be surprised and to surprise others. Just as physical exercise promotes the health of joints and muscles, maintaining a childlike surprisability is tantamount to exercising perceptual curiosity. How does one achieve that? One way is by getting genuinely interested, a few times a week, in at least one of the many events, people, facts, or phenomena we encounter on a daily basis. This could include reading about what determines the path of forked lightning in a rainstorm, inquiring about the hobby of a coworker, examining a new application for the smartphone, following up on a particular tweet, or trying to understand the behavior of the stock market. (Good luck with this last one!) It doesn't really matter what the stimulating object is, as long as one remains excited. Similarly, one should be able to surprise others, and indeed oneself, by doing something unpredictable or seemingly out of character. This could manifest itself in the way one dresses, in engaging with social media, or by changing one's habits. The proactive generation of perceptual curiosity in others seems to strengthen one's own. Curious people like to expose themselves to new sensations and to experience new states of mind. A number of studies have shown that curiosity adds to the motivation derived from the perceived value of information. A study published in 2004 further suggested that curious individuals are attracted to similarly curious individuals, even above and beyond the effects of other shared characteristics.

Not surprisingly, when it comes to fostering curiosity, there is a strategy we can learn from Leonardo: to attempt to create a record of those things that either attract our attention or that we would like to explore. This does not mean one needs to engage in the same type of obsessive note-taking that preoccupied Leonardo for most of his life, but one should at least document those phenomena or events that truly stand out. A subsequent examination of the accumulation of such notes may reveal an underlying theme or pattern worthy of one's epistemic curiosity and may encourage a more thorough study that can produce the pleasure of learning.

The neuroscientific and psychological experiments described in chapters 4–6 (and Feynman's story of preening birds) suggest another way to cultivate curiosity, especially in children and students. Educators should frequently ask questions, but they should not provide the answers right away. Instead, they should encourage their students to give the answer themselves, and then to think of ways to test the correctness of their answers. In other words, the goal is to repeatedly train their epistemic curiosity muscles and to enhance their intellectual dexterity.

Note also that bookstores and libraries offer good opportunities to exercise positive diversive curiosity. Next to the specific book you may be interested in, there are always other books that may be equally interesting. Browsing the internet on topics that pop up during a particular search offers a variant of this experience. You shouldn't miss out on following up (at least occasionally) on such associations since they can be extremely rewarding.

A very important point concerning the refinement of curiosity in students came up in my interview with Martin Rees: a good strategy is to follow the curiosity that the students themselves already have and to recruit the enthusiasm it generates to aid in the teaching process. That is, if the students are eager to know about dinosaurs, start with dinosaurs. As the experiments in chapter 6 have shown, curiosity puts our brain in a state in which it absorbs everything in the vicinity of the object of curiosity. French poet Anatole France perceptively wrote, "The whole art of teaching is only the art of awakening the natural curiosity of young minds for the purpose of satisfying it afterwards."

Something that happened to me personally can help to illustrate the concept. When my youngest daughter was in middle school, the students were asked to choose and carry out a science project. The process is familiar to anyone who has had a child in middle school. These assignments are supposed to invigorate epistemic curiosity, but too often they end up being tedious tasks for the parents. When my daughter asked me what I thought a good science project might be, I came up with the idea of measuring the free-fall acceleration using a number of different methods (a pendulum, an inclined plane, dropping something from the roof, etc.). My daughter promptly declared that all of those experiments were extremely boring and that she would think of a topic herself.

A few days later she told me she would like to test which lipstick could endure the largest number of kisses. This suggestion caught me entirely by surprise, since my daughter had to that point never used any lipstick, nor had she shown any interest in lipstick. Seeing my amazement, she was quick to explain that what she really wanted to test was truth in advertising. Apparently, at the time, one company was making the claim that their lipstick was least eroded by kisses, and my daughter wanted to test the legitimacy of that assertion. I was still a bit unsure about how we would actually conduct the experiment, but my daughter already had one idea. She proposed to put on lipstick and kiss a thin sheet of paper at ten different locations; we would weigh the paper before and after the kisses to determine the weight of the lipstick that stuck to the paper. We would then repeat the process for ten different lipstick brands.

This was starting to shape up as a real scientific experiment, but we still had to find a sufficiently accurate analytical balance to weigh the sheets to the required precision. Here my wife, who is a microbiologist, came to the rescue—she had just the right balance in her lab. In fact, my wife suggested a second, independent test. She also had an instrument that could measure the opacity, or optical depth (in biology they term it "optical density"), of transparent sheets of plastic. That is, the apparatus can determine by how much the intensity of a light beam is attenuated in passing through the sheet. The idea was for my daughter to again put her lips with the different lipsticks to such a transparent sheet, and to use the measurement of the optical density as an independent determination of which lipstick suffered the least amount of erosion. I don't think one needs a better proof for the fact that with a little assistance, following up on questions children are truly curious about can lead to serious explorations. In case you are curious, it turned out that the claim made by that particular lipstick company was true. My daughter also won the first prize in the science competition.

Epilogue

IN 1870, MARK TWAIN PUBLISHED A SHORT STORY that was later titled "A Medieval Romance." The intricate plot is set in the year 1222, and it goes something like this: The conniving Lord of Klugenstein is determined to win the succession of his line from his brother, the Duke of Brandenburgh. Their father, on his deathbed, specified that the succession would go to a male heir or, if none existed, to Brandenburgh's daughter if she proved to be of impeccable reputation. To achieve his treacherous goal, Klugenstein raises his own daughter as if she were a son named Conrad. In addition, to further ensure that Brandenburgh's daughter, Lady Constance, will not be the heir, he dispatches a handsome and shrewd nobleman named Count Detzin to seduce her and thus tarnish her character.

When Brandenburgh's health starts deteriorating, young Conrad is summoned to assume "his" duties as the eventual heir. Klugenstein warns Conrad that a strict law declares that if a female heir sits, even for an instant, on the great ducal throne before she has been crowned, her penalty is death.

The plot thickens when, after a few months into Conrad's role as acting heir, Lady Constance falls in love with "him." When Conrad does not reciprocate, to Constance's great dismay, her love turns into a bitter hatred. To make things worse, Lady Constance, who had indeed been secretly seduced by Klugenstein's conspiring envoy, Count Detzin, gives birth to a child. Detzin himself had long since fled the dukedom.

A trial against Constance commences, and Conrad, with great hesitation, has to sit on the ducal throne as acting duke and judge, even though "he" has not yet been crowned. From that premier chair he solemnly addresses Lady Constance: "By the ancient law of the land, except you produce the partner of your guilt and deliver him up to the executioner, you must surely die. Embrace this opportunity—save yourself while yet you may. Name the father of your child."

It is here that a devastating shock hits. Constance, her eyes beaming with anger, points an accusing finger at Conrad and shouts, "Thou art the man!"

The young judge appears to be inescapably trapped. To reveal "his" gender in order to refute Lady Constance's accusation would mean death for having sat on the forbidden throne. Not to reveal it would also bring the death penalty, for seducing his cousin. How can this incredibly knotty conundrum be resolved?

Fully aware that he has generated a curiosity crescendo for his readers, the witty Twain really shines. He intervenes in the text to admit his inability to disentangle the situation! He simply decides to leave the readers with a perpetual uncertainty, an information gap that is never filled. "The truth is," Twain writes, "I have got my hero (or heroine) into such a particularly close place, that I do not see how I am ever going to get him (or her) out of it again—and therefore I will wash my hands of the whole business, and leave that person to get out the best way that offers—or else stay there."

Is there actually a way to save Conrad from the gallows? Even though Twain couldn't think of one and left his readers to pocket their frustration the best they could, I believe there is still hope for the unfortunate Conrad. I'll reveal my finale before the end of this epilogue.

Twain's story, entertaining as it is, demonstrates the power of curiosity in a very simple yet effective way. We are left apprehensive by the disconcerting lack of resolution. Journalist and author Tom Wolfe performed an analogous sleight-of-hand in his best-selling novel *A Man in Full.* He writes about a couple who check into a motel. Then "she produced that little cup from her handbag, and they did that thing with the cup, something he had never heard of in his life." Many readers have since unsuccessfully speculated about what that sexual practice might have been. A few have even boldly sent their suggestions to Wolfe himself. When asked about it, Wolfe admitted that he had simply concocted the phrase to give the reader a vision of some unmentionable perversion but he had nothing specific in mind.

Other writers use complex contrivances to stimulate something more akin to epistemic curiosity, a yearning to analyze further in order to achieve a more profound understanding. A wonderful example is Samuel Beckett's enigmatic play *Waiting for Godot.* In this two-act absurdist work, two elderly men wait for someone named Godot to show up, but he never does. The play has generated a plethora of interpretations, ranging from spiritual (humanity's need for salvation) to Marxist (an embrace of socialistic values in lieu of capitalistic alienation). Others think the drama reflects Beckett's own experiences during World War II in the French Resistance. Beckett himself, however, appears to have wanted to leave his audience painfully disoriented and curious. "The great success of *Waiting for Godot*," he said, "had arisen from a misunderstanding: critic and public alike were busy interpreting in allegorical or symbolic terms a play which strove above all costs to avoid definition." Equivalently, in late nineteenth-century England, when discussing the fact that a lecture given at the Royal Institution by novelist Walter Besant titled "The Art of Fiction" generated considerable interest, author Henry James noted, "It is proof of life and curiosity—curiosity on the part of the brotherhood of novelists, as well as on the part of their readers." He added: "Art lives upon discussion, upon experiment, upon curiosity, upon variety of attempt, upon the exchange of views and the comparison of standpoints."

Curiosity has undergone a striking reevaluation, from being condemned outright as a vice during medieval times to being extolled as a virtue in the modern age. But is curiosity unequivocally morally good and desirable? There is, for instance, one type of curiosity that is rather bizarre and seemingly inexplicable: *morbid* curiosity. Why would scenes of destruction, violence, mutilation, and death be experienced as an alluring attraction? There are no fewer than three clusters of possible psychological explanations (which suggests that the real reason is not fully understood).

One strand of thought, pioneered by the Swiss psychiatrist Carl Jung, proposes that all people possess a dark side, even if it is buried deep within their mind, under layers of morality. According to this view, our macabre urges represent an attempt to relieve the tension generated by the constant suppression of those forbidden desires. A second theory suggests that the arousal of acute horror that accompanies watching the misery of others is intensely cathartic, leaving the onlooker more relaxed once the experience is over. This idea goes all the way back to Aristotle, who believed that tears produce relief. The great philosopher Immanuel Kant subscribed to it too. A third, somewhat related idea, submits that morbid curiosity creates an empathy with the sufferings of others, which in turn encourages positive social interaction. In other words, morbid curiosity is assumed to represent one part of the evolution of the so-called social brain, which led to more complex forms of sociality. Be that as it may, the mere existence of morbid curiosity demonstrates that we should exercise at least some caution before embracing curiosity in all of its manifestations. This caveat also applies to the fact that negative stories in TV news coverage have been found consistently to appeal more to audiences than positive stories.

Which curiosity-related activities do we fret about today? Governmental surveillance of citizens, such as that performed by the National Security Agency and leaked by Edward Snowden, is certainly one pursuit that raises serious concerns, but it's not the only one. Technology has created many other modern versions of the historical act of eavesdropping. (The term *eavesdropping*, by the way, originally referred to people keeping out of sight, under the eaves, to overhear private conversations inside the house. It is interesting that the obsolete law against eavesdropping in the United Kingdom was abolished only in the Criminal Law Act of 1967.) Present day eavesdropping actions include wiretapping and hacking into email, instant messaging, and other forms of private methods of communication. All of these invasions of personal space are illegal unless allowed by a court order. The semiclandestine accumulation of information by giant companies such as Google, Facebook, and Amazon about our shopping habits, our medical needs, our interests, the literature we read, and other data we consider private, and even intimate, is a form of curiosity that many people frown upon, even though tech companies have at least denied giving the NSA access to their servers. Likewise, the harassment of celebrities by paparazzi has been the subject of numerous lawsuits and headlines. Even certain kinds of scientific research, in particular involving human subjects or serious genetic interventions, are regarded as unethical.

Curiosity comes attached to two connotations: good and bad, legitimate and illegitimate, commendable and controversial. The version I have described, discussed, and emphasized in this book is the good, virtuous curiosity that has precipitated and propelled human intellectual evolution. That's the curiosity that drives education, exploration, and all that is exciting and inspiring in our lives. At the same time, we must be fully aware of the negative aspects of curiosity, especially when we are on its receiving end.

There is another question that deserves some contemplation. With the advent of fast search engines, the existence of Wikipedia, and access to information literally at our fingertips, is there a danger that mystery will be lost and that curiosity (the good kind) will dwindle or be altogether quenched? Do YouTube, Twitter, and Wikipedia erode our ability to be amazed? That was at least partially the opinion expressed in an article entitled "Teach Your Children Well: Unhook Them from Technology," which appeared on January 1, 2016, in the *Wall Street Journal.* Such concerns inform the Waldorf education program which is based on ideas originally expressed by Austrian philosopher Rudolf Steiner. The emphasis of this pedagogy is on the role of imagination and practical experiments in the learning process. Accordingly, Waldorf schools do not introduce computer technology before the early teen years. I should stress, however, that here I am interested only in the effects of information and communications technology on curiosity, and not on the educational experience in general.

Since I could see arguments on both sides of this more specific question, I decided to find out what cognitive scientist Jacqueline Gottlieb thought about it. "It can cut both ways," she told me in a Skype conversation. "For instance, I am a very curious, information-seeking person, so I use the internet as a tool, and I find it extremely useful."

Even though I felt precisely the same, I thought that I should try to play the other side. "Yes, but you grew up without these tools, so maybe you already had the opportunity to become curious?" I wondered.

Gottlieb replied, "Maybe. But curiosity largely comes from inside your brain—how motivated you are to learn and how you learn. If you have a high degree of curiosity from the inside, the internet is not going to change that. Maybe the internet will make a difference for those who are not particularly curious from the inside." After a short pause she added, "If the schools will motivate their students to learn, I don't believe that the students' curiosity will be adversely affected by the internet."

I imagine that this topic will continue to occupy educators and psychologists for at least a few years (if not decades) to come. Also, with artificial intelligence potentially becoming more prominent (see, for example, my interview with Martin Rees in chapter 8), this conversation may take an entirely different direction in the future. Irrespective of what the effects of the internet on curiosity in general may be, however, the internet cannot stop the epistemic curiosity that drives scientific progress. Science is propelled by curiosity about questions we don't know the answers to, and those are precisely the questions for which you cannot find the answers on the internet.

I have not forgotten about Twain's "Medieval Romance." Recall that the protagonist has gotten herself into a pickle, where it appears that to save herself from the accusation of fathering the son of Lady Constance, she has to reveal that she is, in fact, a woman, thereby condemning herself to death for having sat on the forbidden ducal throne. How do I propose to save her? Here is a way out: The Lord of Klugenstein could not have counted on Lady Constance becoming pregnant when he sent Count Detzin himself to seduce her. Neither could he have relied on Count Detzin to testify that he did seduce her, since by doing so Detzin would have sealed his own fate. For his devious plan to tarnish Lady Constance's reputation to succeed, the Lord of Klugenstein had to ensure that someone in the ducal palace (maybe a maid or a guard) would secretly witness the surreptitious seduction and would be prepared to testify about it. This witness could save young Conrad, without Conrad having to reveal that she is a woman.

I think everybody would agree that had Twain finished his "Medieval Romance" with this conclusion, the story would have been much less enchanting (in spite of the happier ending). By leaving us everlastingly curious, Twain achieved a hauntingly memorable effect.

The seventeenth-century lawyer and mathematician Pierre de Fermat managed a far more spectacular feat when he laconically wrote in the margin of his copy of the book *Arithmetica*, "I have discovered a truly remarkable proof which this margin is too small to contain." The proof, which Fermat surely did not have, was supposed to be for what has become known as "Fermat's last theorem"—the most celebrated theorem in number theory. Fermat's intriguing note inspired many generations of curiosity-driven mathematicians to toil unsuccessfully to find a general proof. The theorem was finally fully proved by British mathematician Andrew Wiles, and the two papers presenting the proof (one coauthored with mathematician Richard Taylor) were published in 1995. The curiosity sparked by Fermat's marginal note fueled a major mathematical quest that lasted 358 years.

I hope that I have managed to make the case for the fact that a curious person is someone on whom little is lost. In abandoning the dogmatic pretension of knowledge that characterized humanity during the Middle Ages and replacing it with curiosity, we have managed to usher in and inspire a new way of life. They say that curiosity is contagious. If that's true, my advice would be: *Let's turn it into an epidemic.* As Leonardo put it five centuries ago, "Blinding ignorance does mislead us. O! wretched mortals, open your eyes!"

Notes

Chapter 1: Curious

1 *"The Story of an Hour"*: The story was first published in *Vogue* on December 6, 1894, under the title "The Dream of an Hour." Chopin 1894.

2 *As the English essayist Charles Lamb*: In "Valentine's Day," one of the essays in the collection *Essays of Elia*, which appeared in *London Magazine* between 1820 and 1825.

2 *has been dubbed* empathic *curiosity*: Bateson 1973; McEvoy & Plant 2014.

3 *neuroscientist Joseph LeDoux*: LeDoux described many of his results on fear and surprise in two popular books, LeDoux 1998, 2015.

3 *British Canadian psychologist Daniel Berlyne*: Berlyne published a few seminal papers (e.g., Berlyne 1950, 1954a, b, 1978) and an influential book (Berlyne 1960).

4 *Thomas Hobbes dubbed it*: In *Leviathan*, Hobbes wrote, "*Desire*, to know why, and how, CURIOSITY; such as is in no living creature but *Man*; so that Man is distinguished, not only by his Reason; but also by this singular Passion from other *Animals*; in whom the appetite and other pleasures of Sense, by predominance, take away the care of knowing causes; which is a Lust of the mind, that by a perseverance of delight in the continual and indefatigable generation of Knowledge, exceedeth the short vehemence of any carnal Pleasure." Hobbes 1651, part 1, chap. 6, p. 26.

4 *Einstein alluded to when he told*: Einstein wrote it in a letter to Carl Seelig on March 11, 1952. Einstein Archives at Hebrew University, 39–013. Seelig was a Swiss journalist and writer, who published a biography of Einstein (*Albert Einstein und die Schweiz*) in 1952.

5 *the* morbid *curiosity that results*: Zuckerman 1984; Zuckerman & Litle 1985.

5 *A message in a bottle*: A British scientist named George Parker Bidder dropped more than 1,000 such bottles into the sea to study ocean currents.

One bottle was found only 108 years later. See story at www.cnn.com/2015/08/25/europe/uk-germany-message-in-a-bottle/.

5 *Ed Shevlin*: The Fulbright Commission for Summer Language Study even awarded Mr. Shevlin a grant to study in Ireland. See the story at www.nytimes.com/2011/10/23/nyregion/character-study-ed-shevlin.html.

6 *named Shoemaker-Levy 9*: For a description of the event, 20 years later, see Levy 2014.

8 The Anatomy Lesson of Dr. Nicolaes Tulp: The painting even inspired a novel, Siegal 2014.

9 *neuroscientist Irving Biederman*: Biederman & Vessel 2006.

9 *orator and philosopher Cicero*: Marcus Tullius Cicero wrote this passage in book 5, volume 17 of *De Finibus Bonorum et Malorum* (*About the Ends of Goods and Evils*). Cicero, 1994, p. 449. Discussed also in Zuss 2012.

9 *French philosopher Michel Foucault*: This quote is from "The Masked Philosopher," an interview with Christian Delacampagne in *Le Monde*, April 6, 1980. Foucault opted for the mask of anonymity so as not to influence the readers by his "name." The interview appears in Foucault 1997, where some inaccuracies in the original translation were corrected.

10 *"the most relentlessly curious man"*: In Clark 1969, p. 135.

10 *"It has to do with curiosity"*: "The Feynman Series—Curiosity," an interview with Feynman, https://www.youtube.com/watch?v=ImTmGLxPVyM.

11 *Numerous previous attempts to uncover*: A similar point was made by Fritjof Capra in his excellent book *Learning from Leonardo* (Capra 2013, p. 1).

11 *psychologist Mihaly Csikszentmihalyi*: A fascinating discussion based on nearly 100 interviews in Csikszentmihalyi 1996.

12 *At first there was a point of light*: You can see the entire sequence of images of the aftermath of the first fragment impact at hubblesite.org/newscenter/archive/releases/1994/image/a/format/web_print/.

Chapter 2: Curiouser

13 *short sentences of Giorgio Vasari*: In Vasari 1986, p. 91.

14 *"Those who study the ancients"*: Leonardo expressed these sentiments many times, in slightly different ways. For example, in his MS. E, folio 552, he wrote, "My intention is to cite experience first." The quote here appears also in Nuland 2000.

14 *"Though I may not like them"*: In Richter 1970; can also be found online at https//en.wikisource.org/wiki/The_Notebooks_of_Leonardo_Da_Vinci. See also MacCurdy 1958.

14 *Vasari also provides us*: Vasari 1986, p. 91.

15 *even a partial inventory of his library*: The books are listed in Reti 1972. Originally the list appeared in 1968 in *Burlington Magazine* in London.

15 *Paolo Giovio provided us in 1527*: Giovio 1970.

16 *When Pope Leo X heard*: Vasari (1986, p. 116) tells us that Pope Leo X allotted a certain work to Leonardo, who "straightaway began to distill oils and herbs, in order to make the varnish," which provoked the pope's complaint.

16 *"Study the science of art"*: The more complete quote is "To develop a complete mind, Study the science of art; Study the art of science; Learn to see; Realize that everything connects to everything else."

16 The Last Supper: The painting is in the Santa Maria delle Grazie refectory in Milan. A wonderful description of the various elements in the painting can be found in Keele 1983, p. 24. Entire books devoted to *The Last Supper* are Barcilon and Marani 2001; King 2012.

17 The Virgin and Child with St. Anne: Leonardo kept this painting in his possession until his death. One of the best reproductions in print form is in Zöllner 2007. A beautiful description and discussion of the painting can be found in Clark 1960.

17 *Amazed and Curious*: This title was taken from the poem "Tam O'Shanter" by Robert Burns (1759–1796).

17 *Many excellent studies have attempted*: In the 2008 *Dictionary of Scientific Biography* there are superb discussions by Kenneth Keele, Ladislao Reti, Marshall Clagett, Augusto Marinoni, and Cecil Schneer of Leonardo's studies in anatomy and physiology, technology and engineering, mechanics, mathematics, and geology. Gillispie 2008. Extensive expositions also appear in Kemp 2006; Keele 1983; Galluzzi 2006; Capra 2013; and White 2000. Leonardo's studies of the brain are beautifully described in Pevsner 2014.

17 *produced genuinely new discoveries*: A very detailed description can be found in Hart 1961, and in Kenneth Keele's article in Gillespie 2008.

18 *Leonardo's drawings alone*: A beautiful collection is in Bambach 2003.

18 *the actual content of the notebooks*: MacCurdy 1958; Richter 1952.

19 *the "book of Nature"*: Galileo 1960.

19 *psychiatrist Herman Nunberg would say*: In Nunberg 1961, p. 9. Emphasis in original.

19 *demonstrate the powerful interdependence*: A good description is in Ackerman 1969, p. 205.

19 *thought to have originated*: The *Syracuse Post Standard* used a version of this phrase in an article that appeared on March 28, 1911. Apparently the paper was quoting newspaper editor Arthur Brisbane, who in a talk said, "Use a picture. It's worth a thousand words."

19 *"You who think to reveal"*: MacCurdy 1958, p. 100.

20 *Carlo Pedretti, a Leonardo scholar*: A few of his important books are Pedretti 1957, 1964, and 2005. He is also a coeditor of the facsimile edition of Leonardo's drawings in the Windsor Collection, Clark & Pedretti 1968.

21 "*Painting compels the mind*": In *Treatise on Painting*, para. 55. See also Keele 1983, p. 131 for a discussion of Leonardo's scientific methods.

22 *his drawings of water flows*: Windsor Castle, Royal Library, RL 12579r. Beautifully reproduced in Zöllner 2007, p. 525. Discussed in Gombrich 1969, p. 171.

22 *his painting of Ginevra de' Benci*: There are wonderfully detailed reproductions in Zöllner 2007. The painting is in the National Gallery of Art in Washington, DC (Alisa Mellon Bruce Fund, 1967).

23 *through the language of mathematics*: For example, in *Treatise on Painting*, para. 15.

23 *First, there was the geometry*: An excellent discussion is in Keele's article in Gillispie 2008, p. 193.

23 *Concerning the propagation of light*: Manuscript Ashburnham 2038, fol. 6b, Paris, Institut de France.

23 *the four "powers" of nature*: Leonardo discussed these in relation to many topics, ranging from the operation of the human heart and the flight of birds to the flow of water and various machines. For example, in *Madrid Codex I*, 128v. An excellent discussion can be found in Keele 1983, chapter 4. Leonardo also wrote about gravity, for example: "The power of every gravity is extended toward the center of the world." In *Codex Atlanticus*, fol. 246r-a.

23 *For branching systems*: Described in Kemp 2006.

24 *some aspects of curvilinear geometry*: For an entertaining analysis of Leonardo's works in curvilinear geometry see Wills 1985.

24 *"No man who is not a mathematician": Windsor Collection*, fol. 19118v, in MacCurdy 1958, p. 85.

24 *in the macrocosm of the world*: Leonardo 1996, sheet 3B/folio 34r.

25 *"An object offers as much resistance"*: In *Codex Atlanticus*, fol. 281v-a.

25 *James Playfair McMurrich*: McMurrich 1930.

25 *All Thy Heart Lies Open unto Me*: This title was taken from the poem "Now Sleeps the Crimson Petal" by Tennyson (1809–1892).

25 *investigations into the operation of the human heart*: A detailed description and thorough analysis of Leonardo's studies of the heart is in Keele 1952.

25 *Galen concluded that the heart*: Described in Zeldin 1994, p. 194.

28 *represented the aorta by a glass model*: Leonardo attached the bag representing the ventricle to the glass model and squeezed the bag so that water passed through the aortic valve.

28 *the entire concept and mechanism*: Leonardo mistakenly thought that the impulse of cardiac percussion was completely exhausted at the extremities of the body, so he had no understanding of blood circulation.

29 *to explain phenomena through physical*: An excellent description of Leonardo's efforts is in Zubov 1968.

29 *I Have Seen a Curious Child*: The title was taken from the poem "The Excursion" by Wordsworth (1770–1850).

30 *"Wrongly do men blame": Codex Atlanticus*, 154 r.c. A few somewhat different translations of this text exist (e.g., MacCurdy 1958, p. 64).

30 *"Speculators about perpetual motion": Codex Forster*, II, fol. 92v.

30 *"Having wandered some distance"*: In Richter 1883, vol. 2, p. 395.

31 *"this huge world, which exists"*: In Schilpp 1949, "Autobiographical Notes."

32 *He terms this peculiarity "complexity"*: Csikszentmihalyi 1996, chapter 3.

32 *Leonardo was thought by many*: For example Freud 1916; Farrell 1966. An anonymous charge of homosexuality was actually brought against Leonardo in 1476, but was eventually dismissed.

33 *Consider the following list*: Major characteristics of ADHD are described at www.russellbarkley.org/factsheets/adhd-facts.pdf. See also *Diagnostic and Statistical Manual of Mental Disorders (DSM-5)*, 2013, American Psychiatric Association.

34 *Jonna Kuntsi, an ADHD researcher*: Interviewed by the author on October 7, 2014.

34 *Michael Milham, a neuroscientist*: Interviewed by the author on October 30, 2014. The topic of *high intelligence* is reviewed by Jung (2014).

34 *temperament characteristic of novelty seeking*: For example, Wood et al. 2011; Instanes et al. 2013.

34 *related to the level of the neurotransmitter*: See, for example, Paloyelis et al. 2010, 2012; Lynn et al. 2005.

35 *Bradley Collins wrote in his book*: Collins 1997.

35 *Mark Kac distinguishes*: Kac 1985, p. xxv.

Chapter 3: And Curiouser

37 *"a lot of baloney"*: Feynman tells this story in Feynman 1988, p. 55.

38 *who suffer from Parkinson's disease*: For example, experiments described in Lange et al. 1995; Riesen & Schnider 2001.

38 *shapes called flexagons*: Flexagons were discovered by British mathematician Arthur Harold Stone while he was a student at Princeton University in 1939. Together with fellow graduate students Bryant Tuckerman and Richard Feynman and math instructor John Tukey, they formed the "Princeton Flexagon Committee."

39 *in computer science*: He did pioneering work in quantum computing (e.g., Feynman 1985a).

39 *"Jerry, I have an idea"*: Described in Zorthian, J. H., "We Both Admired Leonardo," in Feynman 1995a, p. 49.

39 *"I wanted to convey an emotion"*: In Feynman 1985, p. 261.

39 *Painting compels the mind*: In *Codex Forster* III 44v. Leonardo made an even stronger statement: "The painter contends with and rivals nature." MacCurdy 1958, p. 913.

40 *Art is I Science is We*: Quote from French physiologist Claude Bernard (1813–1878), in *Bulletin of New York Academy of Medicine*, vol. 4 (1928), p. 997.

40 *"I gave up the idea of trying"*: In Feynman 1985, p. 263.

40 *"Poets say science takes away"*: In Feynman et al. 1964, vol. 1, lecture 3, "The Relation of Physics to Other Sciences,"; section 3-4, "Astronomy." Available online at feynmanlectures.caltech.edu.

40 *English romantic poet John Keats*: The quote is from "Lamia," part 2, line 234. The poem was written in 1819 and published in 1820. It can be found online at www.bartleby.com/126/37.html.

40 *"Art is the tree of life"*: In Blake's annotation to the etching *Laocoön*. The text can be found online at www.betatesters.com/penn/laocoon.htm.

41 *"I too can see the stars"*: in Feynman et al. 1964, vol. 1, lecture 3, "The Relation of Physics to Other Sciences," section 3-4.

42 *"The next girl I met"*: Appears also in Feynman 1995a, p. 27.

42 *"He was genuinely interested in drawing"* Zorthian in Sykes 1994, p. 104. Feynman had a reputation of being a womanizer and perhaps even occasionally a sexist. In fact, Caltech archivist Judith Goodstein and Caltech physicist David Goodstein suggested that one should include women as one of the areas Feynman was interested in. These characteristics of Feynman, if true, are reprehensible. This chapter, however, is not meant to be a comprehensive biography of Feynman. It is intended to demonstrate that he undoubtedly was one of the most curious persons ever to have lived. An excellent article that addresses the alleged aspects of his personality that deserve condemnation is Lipman 1999.

42 *Feynman paid me $5.50*: Conversation with the author on November 3, 2014.

44 *"I don't know if I could really explain"*: Kathleen McAlpine-Myers in Sykes 1994, p. 110.

45 *Leonardo scholar Paolo Galluzzi*: Galluzzi gave a talk entitled "The Shadow of Light: Leonardo's Mind by Candlelight," on March 30, 2011, at the Italian Academy in New York. The lecture is online at italianacademy.columbia.edu/event/shadow-light-leonardos-mind-by-candlelight.

46 *These "funny-looking pictures"*: Feynman introduced these diagrams in a small science meeting in the spring of 1948. The story of the diagrams and their use in physics is excellently told in Kaiser 2005. For an insightful description of the connection between the physics and the beauty of ideas about nature, see Wilczek 2015. See also Feynman 1985b.

47 *That theoretical construct agrees*: The most accurate measurement of the magnetic moment of the electron is in Hanneke et al. 2008. For a brief discussion of the result see gabrielse.physics.harvard.edu/gabrielse/resume.html.

47 *He once told physicist Freeman Dyson*: In Gleick 1992, p. 244.

48 *relation of physics to other branches of Science*: Briefly summarized in Feynman et al. 1964, vol. 1, lecture 3.

48 *enzymes, proteins, and DNA*: It is interesting that a concept Feynman introduced for quantum computing (known as the "Feynman gate") is now being realized by integrating DNA and graphene oxide (e.g., Zhou et al. 2015).

49 *Mrs. Robert Weiner wrote*: The correspondence is reproduced in Feynman 2005, pp. 245–48.

49 *"Look how pretty the stars shine!"*: In Feynman 2001, p. 27. Feynman himself attributed the story to Arthur Eddington. However, Eddington was a lifelong Quaker who never married, which gave rise to the speculation that this anecdote may describe Houtermans.

51 *"We need his [Feynman's] input"*: Zorthian, cited, for example, in William W. Coventry's "A Brief History of Lives in Science," online at wcoventryo.tripod.com/id24.htm. Gell-Mann also grumbled that Feynman "spent a great deal of time and energy generating anecdotes about himself."

51 *"People think they're very close"*: Cited in Gleick 1992, "Epilogue."

52 *Galluzzi is convinced*: Conversation with the author on December 11, 2014.

52 *"Why cannot we write the entire"*: In a talk entitled "There's Plenty of Room at the Bottom," given on December 29, 1959, at the annual meeting of the American Physical Society. First published in *Engineering and Science*, 23:5 (February 1960), 22. Online at www.zyvex.com/nanotech/feynman.html. Feynman offered another prize for a rotating electric motor that would be 1/64th-inch cubed. That prize was claimed by William McLellan.

53 *Tom Newman, then a graduate student*: His story is told in an article entitled "Tiny *Tale* Gets Grand," in *Engineering & Science*, January 1986, p. 25.

54 *Joel Yang of the Singapore University*: The experiment is presented in Tan et al. 2014.

54 *the Nano Bible*: See the 2015 news story "World's Smallest Bible Would Fit on the Tip of a Pen," at www.cnn.com/2015/07/06/middleeast/israel-worlds-smallest-bible/.

54 *"So this man who'd been"*: In Sykes 1994, p. 253.

54 *"This dying is boring"*: In Sykes, 1994, p. 254, only the first part of the sentence is given. A slightly different version of Feynman's last words appears in Gleick 1992, p. 438: "I'd hate to die twice. It's so boring." Joan Feynman insisted in a conversation with the author that the version given here is the correct one.

55 *"This gentleman has compiled"*: Cited in Clark 1975, p. 157.

55 *"While I thought that I was learning*: In *Codex Atlanticus*, 252, r.a. The quote is in MacCurdy 1958, p. 65.

Chapter 4: Curious about Curiosity: Information Gap

57 *psychologist Paul Silvia*: Silvia 2012.

57 *Charles Spielberger and Laura Starr*: Spielberger & Starr 1994.

57 *philosopher Daniel Dennett*: Dennett 1991, pp. 21–22.

58 *Celeste Kidd and Benjamin Hayden*: Kidd & Hayden 2015 provide a nice review of some of the issues involved in the definition of curiosity.

60 *cognitive scientist Laura Schulz*: Schulz studied how very little children react to such situations. See, for example, Cook et al. 2011; Muentener et al. 2012; Bonawitz et al. 2011.

61 *For example, as of December 2015*: See https://www.statista.com/statistics/398166/us-instagram-user-age-distribution/.

61 *recall that psychologist Daniel Berlyne*: In addition to his seminal book (Berlyne 1960), Berlyne wrote a series of very influential papers. For example, on *interest* 1949, on *novelty* (1950), on *perceptual curiosity* (1957), and on *complexity and novelty* (1958). For an article on *specific curiosity* see, Day 1971.

62 *who was also a competent pianist*: A very nice obituary of Berlyne is Konečni 1978. See also www.psych.utoronto.ca/users/furedy/daniel_berlyne.htm.

62 *in a corner nursing a gin and tonic*: in Day 1977.

63 *wrote in his 1978 obituary*: Konečni 1978.

64 *James proposed in the late*: William James was a philosophical giant who helped lay the foundations for many twentieth-century ideas. His work in psychology is summarized in James 1890. His discussion of scientific curiosity is in volume 2. He distinguished between scientific curiosity and the emotional mixture of excitement and anxiety that is associated with exploring novelty. In modern terms this distinction could correspond to the difference between epistemic curiosity and perhaps a blend of perceptual and diversive curiosity.

64 *psychologist George Loewenstein*: The article by Loewenstein (1994) has inspired much of the modern research on curiosity.

64 *investigate and seek new insights*: The relationship between knowledge and curiosity had previously been investigated, e.g., by Jones 1979; Loewenstein et al. 1992.

64 *the information-gap theory identifies* uncertainty: For the more mathematically inclined, the uncertainty is quantified through the entropy that is given by $-\Sigma_{i=1}^{n} p_i \log_2 p_i$, where p_i denotes the probability of outcome *i*.

65 *A body of research in psychology*: For example, Litman & Jimerson 2004; Kang et al. 2009. See also Deci & Ryan 2000 about human needs.

65 *refer to as the* feeling-of-knowing: Loewenstein 1994; Loewenstein et al. 1992; Eysenck 1979; Litman et al. 2005; Hart 1965.

66 *easy to see how* specific *curiosity*: See, for example, Silvia 2006, for a discussion.

66 *In any murder mystery*: The reader is guided from a state of high uncertainty to low uncertainty. See discussion in Gottlieb et al. 2013.

66 *In a study by Cornell psychologists*: Emberson et al. 2010.

67 *requires a knowledge-based mechanism*: An excellent discussion is in Gottl ieb et al. 2013. Basically, one has to accommodate novel information into one's established picture of the world. See also Beswick 1971.

68 *many experiments on exploratory behavior*: See, for instance, Litman 2005; Kashdan & Silvia 2009, (chapter 34); Spielberger & Starr 1994.

68 *the students identified as "curious"*: Ainley 2007.

69 *psychologists Timothy Wilson*: Wilson et al. 2005.

70 *the term "negative capability"*: Keats coined this phrase in a letter to his brothers on December 21, 1817. The letter appears in Keats 2015. All of Keats's letters to his family and friends are in a free ebook, *Letters of John Keats to His Family and Friends*, edited by Sidney Colain.

70 *Roberto Unger, who applied it*: In Unger 2004, p. 279.

70 *John Dewey, who incorporated it*: For example, in Dewey 2005, p. 33.

71 *In Plato's Socratic dialogue*: Online at classics.mit.edu/Plato/meno.html. Discussed in Inan 2012, p. 16.

71 *during a news briefing held in February 2002*: Can be seen on YouTube at: ttps://www.youtube.com/watch?veqGiPelOikQuk.

72 *Foot in Mouth prize*: The British Plain English Campaign annually hands out the prize.

72 *looks like an inverted U*: Berlyne 1970, 1971; Sluckin et al. 1980. Discussed, for example, in Silvia 2006; Edwards 1999, pp. 399–402; and Lawrence

& Nohria 2002, pp. 109–14. A more popular extensive discussion can be found in Leslie 2014.

73 *dates all the way back to Wilhelm Wundt*: Wundt (1832–1920) is sometimes called the "father of experimental psychology." His curve is presented in Wundt 1874.

74 *Berlyne proposed that Wundt's curve*: Berlyne 1971.

75 *a positive reward system*: As we shall discuss later, there is evidence that curiosity activates the dopaminergic system, which is the main reward circuit in the brain (e.g., Redgrave et al. 2008; Bromberg-Martin & Hikosaka 2009).

76 *in the case of the emotion of fear*: LeDoux 2015.

77 *fail to explain the common inverted-U pattern*: This was the conclusion in Silvia 2006.

Chapter 5: Curious about Curiosity: Intrinsic Love of Knowledge

79 *curiosity may provide its own reward*: Discussed, for example, in Ryan & Deci 2000; Silvia 2012; and Kashdan 2004.

79 *Charles Spielberger and Laura Starr proposed*: Spielberger & Starr 1994.

81 *psychologist Jordan Litman*: Litman 2005. Litman continued to examine the hypothesis of I-type and D-type curiosity in a series of experiments and studies, such as Litman & Silvia 2006; Litman & Mussel 2013; Piotrowski et al. 2014.

82 *curiosity comprising a family of mechanisms*: This is beautifully presented in a 2015 proposal entitled "Understanding Curiosity: Behavioral, Computational and Neuronal Mechanisms," which the authors kindly provided. I conducted interviews with Gottlieb on August 27, 2014 and on January 20, 2016, and with Celeste Kidd on June 2, 2015. See also Risko et al. 2012.

82 *trait labeled "openness to experience"*: Explained, for instance, in McCrae & John 1992.

82 *"Big Five" dimensions*: They appear in almost every psychology textbook. See Schacter et al. 2014. One of the original versions is Costa & McCrae 1992. Since then a number of updated versions appeared, such as NEO Five-Factor Inventory-3, published in 2010.

83 *even in the absence of any financial*: See Oudeyer & Kaplan 2007 on intrinsic motivation.

84 *Neuroscientist Jacqueline Gottlieb*: The experiments and their results are described in Baranes et al. 2014. The general question of autonomous exploration is presented in Gottlieb et al. 2013.

85 *"knowledge-based intrinsic motivation"*: The role of intrinsic motivation in general is thought to be allowing the development of a repertoire of skills. Both knowledge-based and competence-based intrinsic motivation are discussed in Mirolli & Baldassarre 2013, p. 49.

87 *"to figure out how children"*: Nicely presented by Laura Schulz in her TED talk, "The Surprisingly Logical Minds of Babies" https://www.ted.com/talks/laura_schulz_the_surprisingly_logical_minds_of_babies?Language=en. Also, conversation with the author on June 25, 2012.

88 *psychologist Elizabeth Spelke*: A wonderful interview with Spelke appeared in the *New York Times* (Angier 2012). Also, conversation with the author in June 2012.

88 *an innate sense of number*: McCrink & Spelke 2016.

88 *the geometry of the space*: Lee et al. 2012; Winkler-Rhoades et al. 2013.

88 *Spelke and colleagues*: For example, Kinzler et al. 2012; Shutts et al. 2011.

89 *how do children select the subjects*: An extensive experiment designed to understand the early development of the mind is under way at the "Babylab" at Birkbeck, University of London, where they monitor the brain and behavior of babies over a period of two and a half years. Described in Geddes 2015.

89 *Celeste Kidd and her collaborators*: Kidd et al. 2012. Also, a conversation with the author on June 2, 2015.

90 *in a simple jack-in-the-box-type*: Schulz & Bonawitz 2007.

90 *curiosity in children is often related*: For instance, Gweon & Schulz 2011. See also Schulz 2012. Experiments by Azzurra Ruggeri and collaborators have shown that even young children adopt strategies of inquiry that increase the efficiency of information gain. Ruggeri and Lombrozo 2015.

91 *the causal relationships that govern*: One of the early modern studies of motivation was by White 1959. An excellent description of the evolutionary drive to construct representations of the causal structure is Gopnik 2000.

91 *the researchers asked children to scrutinize*: Baraff Bonawitz et al. 2012.

93 *Experiments with 1,356 men*: For example, Giambra et al. 1992; Zuckerman et al. 1978.

Chapter 6: Curious about Curiosity: Neuroscience

95 *Functional magnetic resonance imaging*: For a description of the technique, see, for instance, www.ndcn.ox.ac.uk/divisions/fmrib/what-is-fmri/introduction-to-fmri.

95 *taking snapshots of the changes in the blood flow*: This is professionally termed the "hemodynamic response."

95 *In a seminal investigation in 2009*: Kang et al. 2009.

97 *on anticipation of rewarding stimuli*: It was found, for example, that there is an increased functional connectivity between the prefrontal cortex and the reward system in people who are pathological gamblers (e.g., Koehler et al. 2013).

99 *a strengthening of memory in response*: A number of other studies have also shown that motivational states associated with the anticipation of reward (which curiosity triggers) can enhance memory. For instance, Wittman et al. 2011; Shohamy & Adcock 2010; Murayama & Kuhbandner 2011.

100 *Cognitive scientist Marieke Jepma*: A conversation with the author took place on February 4, 2016. The results of her study were published in Jepma et al. 2012.

100 *perceptual curiosity activated brain regions*: The anterior cingulate cortex and the anterior insula. More on the role of the anterior angulate cortex in conflict situations can be found, for example, in van Veen et al. 2001.

100 *activated known reward circuits*: Striatal regions such as the left caudate, putamen, and nucleus accumbens. A good description of the reward mechanisms is in Cohen & Blum 2002.

102 *cognitive scientists Gottlieb, Kidd, and Oudeyer*: Nicely summarized in a proposal entitled "Understanding Curiosity: Behavioral, Computational and Neuronal Mechanisms," kindly provided to the author by J. Gottlieb.

105 *the questions that neuroscientists Matthias Gruber*: Gruber et al. 2014.

107 *Gruber speculated*: Interview with Lecia Bushak in *Medical Daily*, October 2, 2014, www.medicaldaily.com/how-curiosity-enhances-brain-and-stimulates-reward-system-improve-learning-and-memory-306121.

107 *psychologists Brian Anderson and Steven Yantis*: Anderson & Yantis 2013.

108 *neuroscientists Tommy Blanchard*: Blanchard et al. 2015. See also Stalnaker et al. 2015 for a critical examination of the proposed roles for the orbitofrontal cortex.

110 *Joel Voss and his collaborators*: Voss et al. 2011.

111 *is more akin to* traveling waves: In this case the strength of the wave at any given point changes with time (Alexander et al. 2015).

112 *"The Reproducibility Project: Psychology"*: Open Science Collaboration 2015.

112 *a more recent study raised questions*: Gilbert et al. 2016. These researchers claim that their analysis "completely invalidates" the conclusions of the Reproducibility Project. However, Anderson et al. 2016 countered that the reanalysis of Gilbert et al. depends on selective assumptions. Another statistical reevaluation is Etz & Vanderkerckhove 2016.

113 *Frederic Kaplan and Pierre-Yves Oudeyer*: Kaplan & Oudeyer 2007.

115 *Ido Tavor and Saad Jbabdi*: Tavor et al. 2016.

115 *these new insights do not mean*: Some advances have also been made at the molecular level. Scientists discovered that increasing the protein neural calcium sensor-1 in the dentate gyrus of mice enhances exploratory behavior and memory (e.g., Saab et al. 2009). Ornithologists found that variants of the protein-codein gene DRD4 can cause a strong exploratory behavior in a songbird (e.g., Fidler et al. 2007).

116 *Sarnoff Mednick suggested*: Discussed in Kahneman 2011, pp. 67–70.

Chapter 7: A Brief Account of the Rise of Human Curiosity

117 *a few simple facts concerning the human brain*: There are many excellent books about the brain and the mind at a popular science level. A few examples are Eagleman 2015 and Carter 2014 about the structure of the brain; Pinker 1997 about how the mind works. Gregory 1987 is an extensive compilation about concepts related to the brain and mind. Very concise introductions are O'Shea 2005; Encyclopedia Britannica 2008.

118 *Brazilian researcher Suzana Herculano-Houzel*: These are a few of the papers describing her work: Herculano-Houzel, 2010, 2011, 2012a; Herculano-Houzel & Lent 2005; Herculano-Houzel et al. 2007, 2014. For a popular, comprehensive account of brain sizes, number of neurons, and scaling laws, see Herculano-Houzel 2016.

119 *Jon Kaas, a neuroscientist*: Herculano-Houzel et al. 2007.

120 *but about fifty times larger*: The mass as a function of the number of neurons is a power law with an exponent of 1.7.

120 *Gerhard Roth and Ursula Dicke*: Roth & Dicke 2005. They measured intelligence through behavioral complexity. The researchers found that intelligence correlated also with the speed of neuronal activity, which is expected to increase the more densely packed the neurons are.

120 *Daniel Povinelli and Sarah Dunphy-Lelii*: Povinelli & Dunphy-Lelii 2001.

121 *identified the specific area of the brain*: Wang et al. 2015.

122 *how many hours per day*: Detailed time-budget models are in Lehmann et al. 2008.

123 *"It is brains or brawn"*: Explained in Fonseca-Azevedo & Herculano-Houzel 2012, and described in popular terms in Herculano-Houzel 2016.

125 *Donald Johanson discovered*: Lucy's story is told in detail in Johanson and Wong 2009; Johanson and Edy 1981. Many other books incorporate the Lucy discovery and its implications, for example, Tomkins 1998; Mlodinow 2015; Stringer 2011.

127 *the species known as Homo erectus*: Discussed in all texts on human evolution. See, for example, Steudel-Numbers 2006; Van Arsdale 2013.

127 *That phenomenal change occurred*: Bailey & Geary 2009; Coqueugniot et al. 2004; and extensively discussed in Herculano-Houzel 2016.

128 *In his 2009 book*, Catching Fire: Wrangham 2009.

128 *an even more plausible hypothesis*: Aiello & Wheeler 1995 argued that hominins started at some point to consume more energy for the operation of the brain than for the gut, keeping the total consumption rate roughly constant. See also Isler & van Schaik 2009.

128 *controlled use of fire*: Bellomo 1994; Berna et al. 2012; Gowlett et al. 1981.

129 *found in a hearth-like pattern*: Goren-Inbar et al. 2004.

130 *not all researchers agree with the speculation*: C. Loring Brace suggests that fire has been systematically used for cooking for less than 200,000 years. See also discussion in Dunbar 2014 and a brief description in Gibbons 2007.

130 *evolutionary psychologist Robin Dunbar*: Dunbar 2014.

131 *what we identify as the unique human language*: There is a wide range of views on the origin and evolution of the human language. Reviews can be found in Carstairs-McCarthy 2001; Tallerman & Gibson 2012. Specific theories are discussed in Jungers et al. 2003; Deacon 1995. The potential role of the gene FOXP2 is discussed in Enard et al. 2002. The interaction between theoretical linguistics and cognitive neuroscience is discussed in Moro 2008.

131 *a long, evolutionary, Darwinian-like*: This view is held by many present-day scholars and is beautifully and engagingly described in Pinker 1994. Pinker brilliantly presents language as an instinct.

131 *a rather sudden mutation*: This was suggested by the influential linguist Noam Chomsky. See, for example, Chomsky 1988, 1991, 2011. Chomsky argues that the human brain is equipped with a hard-wired universal grammar.

131 *Dunbar suggests*: Dunbar 1996, 2014.

131 *psychologist Elizabeth Spelke*: Angier 2012.

132 *"atlas of the brain"*: A fantastic video showing the work of researcher Jack Gallant and his collaborators can be seen at https://www.youtube.com/watch?v=k6lnJkx5aDQ.

132 *American anthropologist Roy Rappaport*: In Rappaport 1999.

132 *British anthropologist Camilla Power*: Power 2000.

132 *the Blombos Cave*: Henshelwood et al. 2011.

133 *First Agricultural Revolution*: For brief, recent, original, popular accounts of the history of human civilization see Harari 2015; Mlodinow 2015.

133 *the celebrated scientific revolution*: Two classic texts on scientific revolutions and the paradigm shifts associated with them are Kuhn 1962; Cohen 1985. A more recent perspective is Wootton 2015.

Chapter 8: Curious Minds

135 *"The important thing"*: From the memoirs of editor William Miller, quoted in *Life* magazine, May 2, 1955.

135 *physicist Freeman Dyson*: The *New York Times* ran a profile of Dyson entitled "The Civil Heretic" (by Nicholas Dawidoff) on March 25, 2009. A biography of Dyson is Schewe 2013.

136 *In the summer of 2014*: The interview was on July 30, 2014, and it followed an email exchange.

137 The Scientist as Rebel: Dyson 2006, p. 7.

137 *astronaut and polymath Story Musgrave: Air & Space Magazine* published an interview (by Diane Tedeschi) with Story Musgrave in August 2010. The piece was entitled "Veteran Astronaut Story Musgrave: The Only Person to Fly on All Five Space Shuttle Orbiters."

138 *I talked to Musgrave again*: The interview was on August 7, 2014.

139 *polymath, Noam Chomsky*: There are quite a few books on Chomsky and his ideas, such as Harman 1974; d'Agostino 1986; Otero 1994; and one I found especially helpful, McGilvray 2005.

139 *more than 100 books*: His latest book, *Who Rules the World?*, appeared on May 10, 2016.

140 *Chomsky wrote to me*: The email exchange took place on July 6, 2014.

140 *Stanislas Dehaene and his colleagues*: Wang et al. 2015.

140 *Fabiola Gianotti concentrated primarily*: The interview was on September 24, 2015. *Forbes* magazine listed Gianotti in both 2015 and 2016 as one of "the world's 100 most powerful women."

141 *the "God particle"*: The name "God particle" was coined by physicist Leon Lederman, but even Peter Higgs, the person after whom the Higgs boson is called, dislikes it. The discovery of the Higgs boson after forty years of searching has been one of the most important milestones in science in decades. The discovery has been beautifully documented in Carroll 2012; Randall 2013; and the documentary film *Particle Fever*, which was produced by Mark Levinson, David Kaplan, Andrea Miller, Carla Solomon, and Wendy Sax.

144 *Martin Rees*: I interviewed Lord Rees on October 25, 2015. His popular science books include *Our Final Hour, Just Six Numbers, Before the Beginning.*

145 *the challenges and risks that humanity*: In a TED talk Rees described cosmology and what he regards as challenges for humanity in the coming century: www.ted.com/talks/martin_rees_asks_is_this_our_final_century. He also explained these risks in Rees 2003.

148 *Brian May*: The interview took place on November 19, 2015. A brief biography of May can be found at http://brianmay.com/brian/blog.html.

148 *Victorian stereophotography*: An article about May's passion can be found at www.theguardian.com/artanddesign/2014/oct/20/brian-may-stereo-victorian-3d-photos-tate-britain-queen.

151 *I reviewed for him the upcoming searches*: See Livio & Silk 2016 for a brief description of these efforts.

152 *it is difficult to define*: For example, James Flynn, a political science researcher from New Zealand, showed that intelligence scores were changing considerably from generation to generation, and tables of norms had to be frequently changed. Flynn 1984, 1987; Neisser 1998.

152 *Marilyn vos Savant*: The interview was conducted through email on September 3, 2015. A number of newspapers and magazines published articles

about vos Savant. For example, "Meet the World's Smartest Person," by Mary T. Schmich, appeared in the *Chicago Tribune*, September 29, 1985. Also "Is a High IQ a Burden As Much As a Blessing?" by Sam Knight, in the *Financial Times* (April 10, 2009), can be found at www2.sunysuffolk.edu.kasiuka/materials/54/savant.pdf.

155 *As philosopher Martin Heidegger*: Heidegger continued, "Those who idolize 'facts' never notice that their idols only shine in borrowed light," Heidegger 2000, p. 307.

155 *John "Jack" Horner*: The interview took place on September 3, 2015. An intellectual biography of Horner is at mtprof.msun.edeu/Spr2004/horner.html. His 2011 TED talk is at www.ted.com/talks/jack_horner_shape_shifting_dinosaurs?language=en.

155 *"I was likely to find dinosaur bones"*: Randall 2015 presents an original speculation that connects mass extinctions on Earth to the nature of dark matter.

159 *Vik Muniz*: Muniz 2005, p. 12.

159 *Muniz is currently based*: The interview took place on February 17, 2016. See La Force 2016 for an article on Muniz.

159 *the film* Waste Land *documented*: The official trailer can be seen at https://www.youtube.com/watch?v=sNIwh8vT2NU.

163 *Samuel Johnson wrote in 1751*: In *The Rambler*, no. 103, March 12, 1751, online at the Electronic Text Center, University of Virginia Library.

164 *Ilya Prigogine*: See, for instance, the obituary for Prigogine (Petrosky 2003).

165 *in an interview with* Quanta Magazine: Lin 2014.

Chapter 9: Why Curiosity?

167 *In fact, what he said*: Casanova 1922.

167 *have led to a rich and sophisticated*: A few recent books address several aspects of curiosity. Ball 2013 discusses in particular the emergence of modern science. Manguel 2015 examines curiosity from the perspective of several thinkers, such as Dante, David Hume, and Lewis Carroll. Leslie 2014 advocates cultivation of curiosity in view of what he regards as a danger posed by the internet. Grazer and Fishman 2015 describe Grazer's personal experiences that led to the production of celebrated movies and TV shows.

168 *Minnesota Study of Twins Reared Apart*: Excellently described in Bouchard et al. 1990. For a general background on heredity and environmental influences see Bouchard 1998; Plomin 1999.

169 *In 2004, Bouchard reviewed the results*: Bouchard 2004.

169 *Is the strong genetic influence*: Interestingly discussed in Asbury & Plomin 2013.

171 *there were entire periods in human history*: A superb description of the history of "wonder" is in Daston & Park 1998. An interesting discussion can be found in Goodman 1984.

172 *the pseudonym James Bridie*: His real name was Osborne Henry Mavor. He was a medical doctor who served in World War I. The quote is from his play *Mr. Bolfry.*

172 *traditionally known as "Lot's Wife."*: An image of the pillar at Mount Sodom, Israel, can be seen in the Wikipedia article "Lot's Wife."

172 *"Be not curious"*: Ecclesiastes 3:23 (King James Version).

172 *St. Bernard of Clairvaux*: For a comprehensive discussion of the preoccupation with curiosity in early modern France (and Germany), see Kenny 2004.

174 *attitude toward curiosity started to change*: The transformation is beautifully discussed in Kenny 2004; Blumenberg 1982; Ball 2013; and expertly summarized in Daston 2005. Hannam 2011 argues that maybe even medieval times were not as dark as usually described.

174 *The first person to recognize curiosity*: Another fascinating summary of the changing attitude toward curiosity is in Zeldin 1994, chapter 11. For a relatively recent biography of Descartes see Grayling 2005.

174 *Thomas Browne*: A clear and witty description of Browne's life and work is Aldersey-Williams 2015.

174 *Alexander von Humboldt*: Two interesting biographies of Humboldt are Helferich 2004 and McCrory 2010.

175 *One of his biographers wrote*: De Terra 1955.

175 *his multivolume work* Cosmos: Von Humboldt 1997.

175 *Theodor Zeldin*: Zeldin 1994, p. 198.

176 *Brothers Grimm*: An interesting discussion on curiosity in fairy tales in Rigol 1994.

176 *first appeared in print*: In 1598, in Ben Jonson's play *Every Man in His Humour.* Also in Shakespeare's *Much Ado about Nothing.*

177 *around the end of the nineteenth century*: First appeared in print in an 1873 *Handbook of Proverbs* by James Allan Mair, a version of which can be found on Amazon.com.

177 *the* Degenerate Art *exhibition*: In 2014 the Neue Galerie in New York organized a special exhibition that brought together works of art from the 1937 exhibition, along with photos, films, and documents. The exhibition's catalog is Peters 2014.

179 *Malala Yousafzai*: Malala's story is told in Yousafzai & Lamb 2013.

180 *"Being interested in one's work"*: Zeldin 1994, p. 191.

181 *Vladimir Nabokov*: Nabokov 1990, p. 46.

181 *Irish novelist James Stephens*: Stephens 1912, p. 9.

181 *Jocelyn Richard and Kent Berridge*: Richard & Berridge 2011.

182 *why he thought birds did that*: Feynman 1988, p. 14.

184 *one should be able to surprise others*: As part of his advice on how to foster a creative life, Csikszentmihalyi (1996, p. 347) also suggests surprising and being surprised.

184 *curiosity adds to the motivation*: For example, Rossing & Long 1981.

184 *curious individuals are attracted*: See, for instance, Kashdan & Roberts 2004. The relationship between attachment and curiosity was also studied by Mikulincer 1997.

Epilogue

189 *Mark Twain published a short story*: It was first published under the title "An Awful Terrible Medieval Romance," in the *Buffalo Express*, on January 1, 1870. In 1875 it appeared with the title "A Medieval Romance," in Mark Twain's *Sketches, New and Old.*

190 *Twain's story, entertaining as it is*: For analysis and interpretation see Baldanza 1961; Wilson 1987.

190 *his best-selling novel* A Man in Full: Wolfe 1998.

191 *"that thing with the cup"*: Mentioned also in *The Bonfire of the Vanities* and a nonfiction essay in *Hooking up.*

191 *a plethora of interpretations*: The bafflement and "puzzlement" the play has generated is nicely expressed in Atkinson 1956.

191 *author Henry James*: James 1884.

192 *Curiosity has undergone a striking*: The history of curiosity from the late

seventeenth to early nineteenth centuries is marvelously described in the wide-ranging study by Benedict (2001). A beautifully written, concise compendium of various emotions (including curiosity) and human responses is Watts Smith 2015.

192 morbid *curiosity*: Discussed and quantified with a sensation-seeking scale in Zuckerman 1984; Zuckerman & Litle 1985.

192 *Swiss psychiatrist Carl Jung*: Chapter 2 of Jung 1951.

192 *misery of others is intensely cathartic*: A view suggested by Aristotle, who said that humans "enjoy contemplating the most precise images of things whose sight is painful to us," cited in O'Connor 2014. See also Zuckerman & Litle 1985; Kant 2006.

192 *negative stories in TV news*: See Egan et al. 2005 for a cross-cultural study.

192 *performed by the National Security Agency*: Much of the material leaked by Snowden has been published by *The Guardian* in Britain and by *The Washington Post*. National Public Radio posted a short piece presenting the main facts: http://www.mpr.org/sections/parallels/2013/10/23/240239062/five-things-to-know-about-the-nsas-surveillance-activities.

196 *Pierre de Fermat*: The fantastic story of Fermat's last theorem is told in Singh 1997; Aczel 1997.

Bibliography

Ackerman, J. 1969. "Concluding Remarks: Science and Art in the Work of Leonardo," in *Leonardo's Legacy: An International Symposium*, ed. C. O. O'Malley (Berkeley: University of California Press).

Aczel, A. D. 1997. *Fermat's Last Theorem: Unlocking the Secret of an Ancient Mathematical Problem* (New York: Viking).

Aiello, L. C. & Wheeler, P. 1995. "The Expensive Tissue Hypothesis: The Brain and the Digestive System in Human Evolution," *Current Anthropology*, 36, 199.

Ainley, M. 2007. "Being and Feeling Interested: Transient State, Mood, and Disposition," in *Emotion in Education*, ed. P. A. Schutz & R. Pekrun (Burlington, MA: Academic Press).

Aldersey-Williams, H. 2015. *In Search of Sir Thomas Browne: The Life and Afterlife of the Seventeenth Century's Most Inquiring Mind* (New York: Norton).

Alexander, D. M., Trengove, C., & van Leeuven, C. 2015. "Donders Is Dead: Cortical Traveling Waves and the Limits of Mental Chronometry in Cognitive Neuroscience," *Cog. Process.*, 16(4), 365.

Anderson, B. A. & Yantis, S. 2013. "Persistence of Value-Driven Attentional Capture," *J. Exp. Psychol. Hum. Percept. Perform*, 39(1), 6.

Anderson, C. J., et al. 2016. "Response to Comment on 'Estimating the Reproducibility of Psychological Science,'" *Science*, 351, 1037b.

Angier, N. 2012. "Insights from the Youngest Minds," May 1, *New York Times*, www.nytimes/com/2012/05/01/science/insights/in-human-knowledge-from-the-minds-of-babes.html?_r=0.

Asbury, K. & Plomin, R. 2013. *G Is for Genes: The Impact of Genetics on Education and Achievement* (Hoboken, NJ: Wiley-Blackwell).

Atkinson, B. 1956. "Beckett's 'Waiting for Godot,'" *New York Times*, April 20, https://www.nytimes.com/books/97/08/03/reviews/beckett-godot.html.

Bailey, D. & Geary, D. 2009. "Human Brain Evolution," *Human Nature*, 20, 67.

Baldanza, F. 1961, *Mark Twain: An Introduction and Interpretation* (New York: Barnes & Noble).

Ball, P. 2013. *Curiosity: How Science Became Interested in Everything* (Chicago: University of Chicago Press).

Bambach, C. C. 2003. *Leonardo da Vinci: Master Draftsman* (New York: Metropolitan Museum of Art).

Baraff Bonawitz, E., van Schijndel, T. J. P., Friel, D., & Schulz, L. 2012. "Children Balance Theories and Evidence in Exploration, Explanation, and Learning," *Cognitive Psychology*, 64, 215.

Baranes, A. F., Oudeyer, P.-Y., & Gottlieb, J. 2014. "The Effects of Task Difficulty, Novelty and the Size of the Search Space on Intrinsically Motivated Exploration," *Front. Neurosci.*, 8, 317.

Barcilon, P. B. & Marani, P. C. 2001. *Leonardo: The Last Supper*, trans. Harlow Tighe (Chicago: University of Chicago Press).

Bateson, G. 1973. *Steps to an Ecology of Mind* (London: Paladin).

Bellomo, R. V., 1994. "Methods of Determining Early Hominid Behavioral Activities Associated with the Controlled Use of Fire at FxJj20 Main, Koobi Fora, Kenya," *Journal of Human Evolution*, 27, 173.

Benedict, B. M. 2001. *Curiosity: A Cultural History of Early Modern Inquiry* (Chicago: University of Chicago Press).

Berlyne, D. E. 1949. "Interest as a Psychological Concept," *British Journal of Psychology*, 39, 184.

Berlyne, D. E., 1950. "Novelty and Curiosity as Determinants of Exploratory Behavior," *British Journal of Psychology*, 41, 68.

Berlyne, D. E. 1954a. "A Theory of Human Curiosity," *British Journal of Psychology*, 45, 180.

Berlyne, D. E. 1954b. "An Experimental Study of Human Curiosity," *British Journal of Psychology*, 45, 256.

Berlyne, D. E. 1957. "Determinants of Human Perceptual Curiosity," *Journal of Experimental Psychology*, 53, 399.

Berlyne, D. E. 1958. "The Influence of Complexity and Novelty in Visual Figures on Orienting Responses," *Journal of Experimental Psychology*, 55, 289.

Berlyne, D. E. 1960. *Conflict, Arousal and Curiosity* (New York: McGraw-Hill).

Berlyne, D. E. 1966. "Curiosity and Exploration," *Science*, 153, 25.

Berlyne, D. E. 1970. "Novelty, Complexity and Hedonic Value," *Perception and Psychophysics*, 8, 279.

Berlyne, D. E. 1971. *Aesthetics and Psychobiology* (New York: Appleton-Century-Crofts).

Berlyne, D. E. 1978. "Curiosity and Learning," *Motivation and Emotion*, 2, 97.

Berna, F., et al. 2012. "Microstrategraphic Evidence of in Sita Fire in the Acheulean Strata of Wonderwerk Cave, Northern Cape Province, South Africa," *Proc. of the Natl. Acad. of Sci.*, USA, 109, E1215.

Beswick, D. G. 1971. "Cognitive Process Theory of Individual Differences in Curiosity," in *Intrinsic Motivation: A New Direction in Education*, ed. H. I. Day, D. E. Berlyne, & D. E. Hunt (Toronto: Holt, Rinehart and Winston).

Biederman, I. & Vessel, E. A. 2006. "Perceptual Pleasure and the Brain," *American Scientist*, 94, 249.

Blanchard, T. C., Hayden, B. Y., & Bromberg-Martin, E. S. 2015. "Orbitofrontal Cortex Uses Distinct Codes for Different Choice Attributes in Decisions Motivated by Curiosity," *Neuron*, 85(3), 602.

Blumenberg, H. 1987. *The Genesis of the Copernican World*, trans. R. M. Wallace (Cambridge, MA: MIT Press).

Bonawitz, E., et al. 2011. "The Double-Edged Sword of Pedagoga: Instruction Limits Spontaneous Exploration and Discovery," *Cognition*, 120, 322.

Bouchard, T. J. 1998. "Genetic and Environmental Influences on Adult Intelligence and Special Mental Abilities," *Human Biology*, 70, 257.

Bouchard, T. J., et al. 1990. "Sources of Human Psychological Differences: The Minnesota Study of Twins Reared Apart," *Science*, 250, 223.

Bouchard Jr., T. J. 2004. "Genetic Influence on Human Psychological Traits," *Current Directions in Psychological Science*, 13(4), 148.

Bromberg-Martin, E. S. & Hikosaka, O. 2009. "Midbrain Dopamine Neuron Signal Preference for Advance Information about Upcoming Rewards," *Neuron*, 63, 119.

Capra, F. 2013. *Learning from Leonardo: Decoding the Notebooks of a Genius* (San Francisco: Berelt-Koehler).

Carroll, S. 2012. *The Particle at the End of the Universe: How the Hunt for the Higgs Boson Leads Us to the Edge of a New World* (New York: Dutton).

Carstairs-McCarthy, A. 2001. "Origins of Language," in *The Handbook of Linguistics*, ed. M. Aromoff & J. Rees-Miller (Oxford: Blackwell).

Carter, R. 2014. *The Human Brain Book*, 2nd edition (New York: DK Publishing).

Casanova, G. 1922. *The Memoirs of Giacomo Casanova Di Seingalt*. Trans. A. Machen (London: The Casanova Society), vol. 7.

Chomsky, N. 1988. *Language and Problems of Knowledge: The Managua Lectures* (Cambridge, MA: MIT Press).

Chomsky, N. 1991. "Linguistics and Cognitive Science: Problems and Mysteries," in *The Chomskyan Turn*, ed. A. Kasher (Oxford: Blackwell).

Chomsky, N. 2011. "Language and Other Cognitive Systems: What Is Special about Language?," *Language Learning and Development*, 7(4), 263.

Chopin, K. 1894. "The Story of an Hour," Kate Chopin International Society, www.katechopin.org/story-hour/.

Cicero. 1994. *Cicero: De Finibus Bonorum et Malorum*, trans. H. Rackham (Cambridge, UK: Cambridge University Press).

Clark, K. 1960. "Leonardo da Vinci: The Virgin with St. Anne," in *Looking at Pictures* (New York: Holt, Rinehard and Winston).

Clark, K. 1969. *Civilisation: A Personal View* (New York: Harper & Row).

Clark, K. 1975. *Leonardo da Vinci: An Account of His Development As An Artist* (London: Penguin Books).

Clark, K. & Pedretti, C. (eds.). 1968. *The Drawings of Leonardo da Vinci in the Collection of Her Majesty the Queen*, 3 vols. (London: Phaidon).

Clayton, M. 2012. "Leonardo's Anatomy Years," *Nature*, 484, 314.

Cohen, I. B. 1985. *Revolution in Science* (Cambridge, MA: Belknap Press of Harvard University Press).

Cohen, J. D. & Blum, K. I. 2002. "Overview: Award and Decision." Introduction to special issue, *Neuron*, 36(2), 193.

Collins, B. 1997. *Leonardo, Psychoanalysis, and Art History: A Critical Study of Psychobiographical Approaches to Leonardo da Vinci* (Evanston, IL: Northwestern University Press).

Cook, C., Goodman, N. D., & Schulz, L. E. 2011. "Where Science Starts: Spontaneous Experiments in Preschoolers' Exploratory Play," *Cognition*, 120, 341.

Coqucugniot, H., Hublen, J.-J., Veillon, F., Honët, F., & Jacob, T. 2004. "Early Brain Growth in Homo Erectus and Implications for Cognitive Ability," *Nature*, 431, 299.

Costa Jr., P. T. & McCrae, R. R. 1992. *Revised NEO Personality Inventory (NEO PI-R) and NEO Five-Factor Inventory (NEO-FFI): Professional Manual* (Odessa, FL: Psychological Assessment Resources).

Csikszentmihalyi, M. 1996. *Creativity: Flow and the Psychology of Discovery and Invention* (New York: Harper Collins).

D'Agostino, F. 1986. *Chomsky's System of Ideas* (Oxford: Oxford University Press).

Daston, L. 2005. "All Curls and Pearls," *London Review of Books*, 27(12), 37.

Daston, L. J. & Park, K. 1998. *Wonders and the Order of Nature 1150–1750* (New York: Zone Books).

Day, H. I. 1971. "The Measurement of Specific Curiosity," in *Intrinsic Motivation: A New Direction in Education*, ed. H. I. Day, D. E. Berlyne, & D. E. Hunt (New York: Holt, Rinehart & Winston).

Day, H. I. 1977. "Daniel Ellis Berlyne (1924–1976)," *Motivation and Emotion*, 1(4), 377.

Deacon, T. W. 1995. *The Symbolic Species: The Coevolution of Language and the Human Brain* (Harmondsworth, UK: Allen Lane).

Deci, E. L. & Ryan, R. M. 2000. "The 'What' and "Why' of Goal Pursuits: Human Needs and the Self-Determination of Behavior," *Psychological Inquiry*, 11(4), 227.

Dennett, D. C. 1991. *Consciousness Explained* (Boston: Little, Brown).

de Terra, H. 1955. *Humboldt* (New York: Knopf).

Dewey, J. 2005. *Art as Experience* (New York: Perigee). Originally published in 1934.

Dunbar, R. 1996. *Grooming, Gossip and the Evolution of Language* (London: Faber and Faber).

Dunbar, R. 2014. *Human Evolution* (London: Pelican).

Dyson, F. 2006. *The Scientist as Rebel* (New York: New York Review of Books).

Dyson, G. 2012. *Turing's Cathedral: The Origins of the Digital Universe* (London: Allen Lane).

Eagleman, D. 2015. *The Brain: The Story of You* (New York: Pantheon).

Edwards, D. C. 1999. *Motivation and Emotion: Evolutionary, Physiological, Cognitive, and Social Influences* (Thousand Oaks, CA: Sage).

Egan, V., et al. 2005. "Sensational Interests, Mating Effort, and Personality: Evidence for Cross-Cultural Validity," *Journal of Individual Differences*, 26(1), 11.

Emberson, L. L., Lupyan, G., Goldstein, M. H., & Spivy, M. J. 2010. "Overheard Cell-Phone Conversations: When Less Speech Is More Distracting," *Psychological Science*, 21(10), 1383.

Enard, W., et al. 2002. "Molecular Evolution of FOXP2, a Gene Involved in Speech and Language," *Nature*, 418, 869.

Encyclopaedia Britannica. 2008. *The Britannica Guide to the Brain: A Guided Tour of the Brain—Mind, Memory and Intelligence* (London: Robinson).

Etz, A. & Vanderkerckhove, J. 2016. "A Bayesian Perspective on the Reproducibility Project: Psychology," *PLoS ONE*, 11(2): e 0149794.

Eysenck, M. W. 1979. "The Feeling of Knowing a Word's Meaning," *British Journal of Psychology*, 70, 243.

Farrell, B. 1966. "On Freud's Study of Leonardo," in *Leonardo da Vinci: Aspects of the Renaissance Genius*, ed. M. Philipson (New York: George Braziller).

Feynman, M. 1995a. *The Art of Richard P. Feynman: Images by a Curious Character* (New York: Routledge).

Feynman, M. (compiler). 1995b. *The Art of Richard P. Feynman: Images by a Curious Character* (Basel: G&B Science).

Feynman, R. P. 1985. "Surely You're Joking Mr. Feynman!," in *Adventures of a Curious Character*, ed. Edward Hutchings (New York: Norton).

Feynman, R. P. 1985a. "Quantum Mechanical Computers," *Optics News*, 11, 11.

Feynman, R. P. 1985b. *QED: The Strange Theory of Light and Matter* (Princeton, NJ: Princeton University Press).

Feynman, R. P. 1988. *What Do You Care What Other People Think? Further Adventures of a Curious Character*, ed. Ralph Leighton (New York: Norton).

Feynman, R. P. 2001. *What Do You Care What Other People Think? Further Adventures of a Curious Character*, as told to Ralph Leighton (New York: Norton).

Feynman, R. P. 2005. *Perfectly Reasonable Deviations (From the Beaten Track)*, ed. M. Feynman, foreword by Timothy Ferris (New York: Basic Books).

Feynman, R. P., Leighton, R. B., & Sands, M. 1964. *Feynman Lectures on Physics* (New York: Addison Wesley).

Fidler, A. E., et al. 2007. "Drd4 Gene Polymorphisms Are Associated with Personality Variation in a Passerine Bird," *Proc. of the Royal Society London B.*, 2 May.

Flynn, J. R. 1984. "The Mean IQ of Americans: Massive Gains 1932 to 1978," *Psychological Bulletin*, 95(1), 29.

Flynn, J. R. 1987. "Massive IQ Gains in 14 Nations: What IQ Tests Really Measure," *Psychological Bulletin*, 101(2), 171.

Fonseca-Azevedo, K. & Herculano-Houzel, S. 2012. "Metabolic Constraint Imposes Tradeoff between Body Size and Number of Brain Neurons in Human Evolution," *PNAS*, 109(45), 18571.

Foucault, M. 1997. *Ethics: Subjectivity and Truth*, ed. Paul Rabinow (New York: New Press).

Freud, S. 1916. *Leonardo da Vinci: A Psychosexual Study of an Infantile Reminiscence*, trans. A. A. Brill (New York: Moffat, Yard).

Galileo, 1960. *The Assayer* [*Il Saggiatore*], in *The Controversy on the Comets of 1618*, trans. S. Drake & C. D. O'Malley (Philadelphia: University of Pennsylvania Press).

Galluzzi, P. (ed.). 2006. *The Mind of Leonardo: The Universal Genius at Work*, trans. C. Frost & J. M. Reifsnyder (Florence: Giventi).

Geddes, L. 2015. "The Big Baby Experiment," *Nature*, 527, 22.

Gerges, F. A. 2016. *ISIS: A History* (Princeton, NJ: Princeton University Press).

Giambra, L. M., Camp, C. J., & Grodsky, A. 1992, "Curiosity and Stimulation Seeking across the Adult Life Span: Cross-Sectional and 6- to 8-Year Longitudinal Findings," *Psychology and Aging*, 7(1), 150.

Gibbons, A. 2007. "Food for Thought: Did the First Cooked Meals Help Fuel the Dramatic Evolutionary Expansion of the Human Brain?" *Science*, 316, 1558.

Gilbert, D. T., King, G., Pettigrew, S., & Wilson, T. D. 2016. "Comment on 'Estimating the Reproducibility of Psychological Science,'" *Science*, 351, 1037.

Gillispie, C. C. (ed.). 2008. *Dictionary of Scientific Biography* (New York: Charles Scribner's Sons).

Giovio, P. 1970. *Leonardo Vincii Vita*, reproduced in J. P. Richter & I. A. Richter, *The Literary Works of Leonardo da Vinci*, 3rd edition, vol. 1 (London: Phaidon).

Gleick, J. 1992. *Genius: The Life and Science of Richard Feynman* (New York: Pantheon).

Gombrich, E. H. 1969. "The Form of Movement in Water and in Air," in *Leonardo's Legacy: An International Symposium*, ed. C. D. O'Malley (Berkeley: University of California Press).

Goodman, N. 1984. *Of Mind and Other Matters* (Cambridge, MA: Harvard University Press).

Gopnik, A. 2000. "Explanation as Orgasm and the Drive for Causal Understanding: The Evolution, Function and Phenomenology of the Theory-Formation System," in F. Keil & R. Wilson (Eds.), *Cognition and Explanation* (Cambridge, MA: MIT Press).

Goren-Inbar, N., Alperson, N., Kislev, M. E., Simcroni, O., Melamed, Y., Ben-Nun, A., & Werker, E. 2004. "Evidence of Hominin Control of Fire at Gesher Benot Ya'aqov, Israel," *Science*, 304(5671), 725.

Gottlieb, J., Oedeyer, P.-Y., Lopes, M., & Baranes, A. 2013. "Information-Seeking, Curiosity, and Attention: Computational and Neural Mechanisms," *Trends in Cognitive Sciences*, 17(11), 585.

Gowlett, J. A. J., et al. 1981. "Early Archaeological Sites, Hominid Remains and Traces of Fire from Chesowanja, Kenya," *Nature*, 294, 125.

Grayling, A. C. 2005. *Descartes: The Life and Times of a Genius* (New York: Walker).

Grazer, B. & Fishman, C. 2015. *A Curious Mind: The Secret to a Bigger Life* (New York: Simon & Schuster).

Gregory, R. L. (ed.). 1987. *The Oxford Companion to the Mind* (Oxford: Oxford University Press).

Gruber, M. J., Gelman, B. D., & Ranganath, C. 2014. "States of Curiosity Modulate Hippocampus-Dependent Learning via the Dopaminergic Circuit," *Neuron*, 84(2), 486.

Gweon, H. & Schulz, L. E. 2011. "16-Month-Olds Rationally Infer Causes of Failed Actions," *Science*, 332, 1524.

Hannam, J. 2011. *The Genesis of Science: How the Christian Middle Ages Launched the Scientific Revolution* (Washington, DC: Regnery).

Hanneke, D., Fogwell, S., & Gabrielse, G. 2008. "New Measurement of the Electron Magnetic Moment and the Fine Structure Constant," *Physical Review Letters*, 100, 120801.

Harari, Y. N. 2015. *Sapiens: A Brief History of Humankind* (New York: Harper Collins).

Harman, G. (ed.). 1974. *On Noam Chomsky: Critical Essays* (New York: Anchor Press).

Hart, I. B. 1961. *The World of Leonardo da Vinci: Man of Science, Engineer and Dreamer of Flight* (New York: Viking).

Hart, J. T. 1965. "Memory and the Feeling-of-Knowing Experience," *Journal of Educational Psychology*, 56, 208.

Heidegger, M. 2000. *Contributions to Philosophy*, trans. P. Emad & K. Maly (Bloomington: Indiana University Press).

Helferich, G. 2004. *Humboldt's Cosmos: Alexander von Humboldt and the Latin American Journey That Changed the Way We See the World* (New York: Gotham Books).

Henshelwood, C. S., et al. 2011. "A 100,000-Year-Old Ochre-Processing Workshop at Blombos Cave, South Africa," *Science*, 334, 219.

Herculano-Houzel, S. 2009. "The Human Brain in Numbers: A Linearly Scaled-Up Primate Brain," *Frontiers in Human Neuroscience*, 3, 31.

Herculano-Houzel, S. 2010. "Coordinated Scaling of Cortical Cerebellar Number of Neurons," *Frontiers in Neuroanatomy*, 4, 12.

Herculano-Houzel, S. 2011. "Not All Brains Are Made the Same: New Views on Brain Scaling in Evolution," *Brain Behav. Evol.*, 78, 22.

Herculano-Houzel, S. 2012a. "Neuronal Scaling Rules for Primate Brains: The Primate Advantage," *Prog. Brain Res.*, 195, 325.

Herculano-Houzel, S. 2012b. "The Remarkable, yet Not Extraordinary, Human Brain as a Scaled-up Primate Brain and Its Associated Cost," *PNAS*, 109 (suppl. 1), 10661.

Herculano-Houzel, S. 2016. *The Human Advantage: A New Understanding of How Our Brain Became Remarkable* (Cambridge, MA: MIT Press).

Herculano-Houzel, S., Collins, L. E., Wong, P., & Kaas, J. H. 2007. "Cellular Scaling Rules for Primate Brains," *Proc. Natl. Acad. Sci. USA*, 104, 3562.

Herculano-Houzel, S. & Lent, R. 2005. "Isotropic Fractionator: A Simple Rapid Method for the Quantification of Total Cell and Neuron Numbers in the Brain," *J. Neurosci.*, 25, 2518.

Herculano-Houzel, S., Manger, P. R., & Kaas, J. H. 2014. "Brain Scaling in Mammalian Brain Evolution as a Consequence of Concerted and Mosaic Changes in Number of Neurons and Average Neuronal Cell Size," *Front. Neuroanat.*, 8, 77.

Hobbes, T. 1651. *Leviathan*, Online Library of Liberty, oll.libertyfund.org /titles/869.

Huron, D. 2006. *Sweet Anticipation: Music and the Psychology of Expectation* (Cambridge, MA: MIT Press).

Inan, I. 2012. *The Philosophy of Curiosity* (New York: Routledge).

Instanes, J. T., Haavik, J., & Halmøy, A. 2013. "Personality Traits and Comorbidity in Adults with ADHD," *Journal of Attention Disorder*, Nov 22.

Isler, K. & van Schaik, C. P. 2009. "The Expensive Brain: A Framework for Explaining Evolutionary Changes in Brain Size," *J. Hum. Evol.*, 57, 392.

James H. 1884. "The Art of Fiction," *Longman's Magazine*, 4 (September), public.wsu.edu/~campbelld/amlit/artfiction.html.

James, W. 1890. *The Principles of Psychology, American Science Series, Advanced Course*, 2 vol. (New York: Holt), https://ebooks.adelaide.edu.au/j/james /william/principles/index.html.

Jepma, M., et al. 2012. "Neural Mechanisms Underlying the Induction and Relief of Perceptual Curiosity," *Frontiers in Behavioral Neuroscience*, 6, 5.

Johanson, D. C. & Edy, M. A. 1981. *Lucy: The Beginning of Humankind* (New York: Simon & Schuster).

Johanson, D. C. & Wong, K. 2009. *Lucy's Legacy: The Quest for Human Origins* (New York: Crown).

Jones, S. 1979. "Curiosity and Knowledge," *Psychological Reports*, 45, 639.

Jung, C. 1959. *Aion: Researchers into the Phenomenology of the Self*, in *The Collected Works of C. G. Jung*, trans. R. F. C. Hull, vol. 9, part 2 (Princeton, NJ: Princeton University Press).

Jung, R. E. 2014. "Evolution, Creativity, Intelligence, and Madness: 'Here Be Dragons,'" *Frontiers in Psychology*, 5, article 784, 1.

Jungers, W. L., et al. 2003. "Hypoglossal Canal Size in Living Hominoids and the Evolution of Human Speech," *Human Biology*, 75, 473.

Kac, M. 1985. *Enigmas of Chance: An Autobiography* (New York: Harper Collins).

Kahneman, D. 2011. *Thinking, Fast and Slow* (New York: Farrar, Straus and Giroux).

Kaiser, D. 2005. "Physics and Feynman's Diagrams," *American Scientist*, 93, 156.

Kandel, E. R. 2012. *The Age of Insight: The Quest to Understand the Unconscious in Art, Mind, and Brain* (New York: Random House).

Kang, M. J., et al. 2009. "The Wick in the Candle of Learning: Epistemic Curiosity Activates Reward Circuitry and Enhances Memory," *Psychol. Sci.*, 20(8), 963.

Kant, I. 2006. *Anthropology from a Pragmatic Point of View*, trans. R. B. Louden (Cambridge, UK: Cambridge University Press).

Kaplan, F. & Oudeyer, P.-Y. 2007. "In Search of the Neural Circuits of Intrinsic Motivation," *Front. Neurosci.*, 1(1), 225.

Kashdan, T. B. 2004. "Curiosity," in *Character Strengths and Virtues*, ed. C. Peterson & M. E. P. Selegman (New York: Oxford University Press).

Kashdan, T. B. & Roberts, J. E. 2004. "Trait and State Curiosity in the Genesis of Intimacy: Differentiation from Related Constructs," *Journal of Social and Clinical Psychology*, 23(6), 792.

Kashdan, T. B. & Silvia, P. J. 2009. "Curiosity and Interest: The Benefits of Thriving on Novelty and Challenge," in *The Oxford Handbook of Positive Psychology*, ed. S. J. Lopez & L. R. Snyder (Oxford: Oxford University Press).

Keats, J. 2015. *Selected Letters*, ed. John Barnard (London: Penguin Classics).

Keele, K. D. 1952. *Leonardo da Vinci on Movement of the Heart and Blood* (London: Harvey and Blythe).

Keele, K. D. 1983. *Leonardo da Vinci's Elements of the Science of Man* (New York: Academic Press).

Kemp, M. 2006. *Seen|Unseen: Art, Science and Intuition from Leonardo to the Hubble Telescope* (Oxford: Oxford University Press).

Kenny, N. 2004. *The Uses of Curiosity in Early Modern France and Germany* (Oxford: Oxford University Press).

Kidd, C. & Hayden, B. Y. 2015. "The Psychology and Neuroscience of Curiosity," *Neuron*, 88 (3) 499.

Kidd, C., Piantadosi, S. T., & Aslin, R. N. 2012. "The Goldilocks Effect: Human Infants Allocate Attention to Visual Sequences That Are Neither Too Simple nor Too Complex," *PLoS ONE* 7(5): e 36399.

King, R. 2012. *Leonardo and The Last Supper* (New York: Walker).

Kinzler, K. D., Shutts, K., & Spelke, E. S. 2012. "Language-Based Social Preferences among Children in South Africa," *Language Learning and Development*, 8, 215.

Koehler, S., Ovadia-Caro, S., van der Meer, E., Villringer, A., Heinz, A., Romanczuk-Seifereth, N., & Margulies, D. S. 2013. "Increased Functional Connectivity between Prefrontal Cortex and Reward System, *PLoS ONE*, 8(12), e84565.

Konečni, V. J. 1978. "Daniel E. Berlyne 1924–1976," *American Journal of Psychology*, 91(1), 133.

Kuhn, T. S. 1962. *The Structure of Scientific Revolutions* (Chicago: University of Chicago Press).

La Force, T. 2016. "Master of Illusions," *Apollo*, 183(639), 46.

Lange, K. W., Tucha, O., Steup, A., Gsell, W., & Naumann, M. 1995. "Subjective Time Estimation in Parkinson's Disease," *J. Neural Transm Suppl.*, 46, 433.

Lawrence, P. R. & Nohria, N. 2002. *Driven: How Human Nature Shapes Our Choices* (San Francisco: Jossey-Bass).

LeDoux, J. 1998. *The Emotional Brain: The Mysterious Underpinnings of Emotional Life* (New York: Simon & Schuster).

LeDoux, J. 2015. *Anxious: Using the Brain to Understand and Treat Fear and Anxiety* (New York: Viking).

Lee, S. A., Winkler-Rhoades, N., & Spelke, E. S. 2012. "Spontaneous Reorientation Is Guided by Perceived Surface Distance," *PLoS ONE*, 7, e51373.

Lehmann, J., Korstjens, A. H., & Dunbar, R. I. M. 2008. "Time and Distribution: A Model of Ape Biogeography," *Ecology, Evolution and Ethology*, 20, 337.

Leonardo da Vinci. 1996. *Codex Leicester: A Masterpiece of Science*, ed. Claire Farago, with introductory essays by Martin Kemp, Owen Gingerich, and Carlo Pedretti (New York: American Museum of Natural History).

Leslie, I. 2014. *Curious: The Desire to Know and Why Your Future Depends on It* (New York: Basic Books)

Levy, D. H. 2014. "Comet Shoemaker-Levy 9:20 years later," *Sky & Telescope*, July 16, www.skyandtelescope.com/astronomy-news/comet-shoemaker-levy-9-20-years-later-07162014/.

Lin, T. 2014. "A 'Rebel' without a Ph.D.," *Quanta Magazine*, March 26, 2014, https://www.quantamagazine.org/20140326-a-rebel-without-a-ph-d/.

Lipman, J. C. 1999. "Finding the Real Feynman," *The Tech*, 119 (10), tech.mit.edu/V119/N10/col10lipman.10c.html.

Litman, J. A. 2005. "Curiosity and the Pleasure of Learning: Wanting and Liking New Information," *Cognition and Emotion*, 19(6), 793.

Litman, J. A. & Jimerson, T. L. 2004. "The Measurement of Curiosity as a Feeling of Deprivation," *Journal of Personality Assessment*, 82(2), 157.

Litman, J. A., Hutchins, T. L., & Russon, R. K. 2005. "Epistemic Curiosity, Feeling-of-Knowing, and Exploratory Behavior," *Condition and Emotion*, 19(4), 559.

Litman, J. & Silvia, P. 2006. "The Latent Structure of Trait Curiosity: Evidence for Interest and Deprivation Curiosity Dimensions," *Journal of Personality Assessment*, 86 (3), 318.

Litman, J. A. & Mussel, P. 2013. "Validity of the Interest- and Deprivation-Type Epistemic Curiosity Model in Germany," *Journal of Individual Differences*, 34(2), 59.

Livio, M. & Silk, J. 2016. "If There Are Aliens Out There, Where Are They?," *Scientific American*, January 6, www.scientificamerican.com/article/if-there-are-aliens-out-there-where-are-they/.

Locke, J. L. 2010. *Eavesdropping: An Intimate History* (Oxford: Oxford University Press).

Loewenstein, G. 1994. "The Psychology of Curiosity: A Review and Reinterpretation," *Psychological Bulletin*, 116(1), 75.

Loewenstein, G., Adler, D., Behrens, D., & Gilles, J. 1992. "Why Pandora Opened the Box: Curiosity Is a Desire for Missing Information," Working paper, Dept. of Social and Decision Sciences (Pittsburgh, PA: Carnegie Mellon University).

Lynn, D. E., et al. 2005. "Temperament and Character Profiles and the Dopamine D_4 Receptor Gene in ADHD," *American Journal of Psychiatry*, 162, 906.

MacCurdy, E. 1958. *The Notebooks of Leonardo da Vinci* (New York: George Braziller).

Manguel, A. 2015. *Curiosity* (New Haven, CT: Yale University Press).

McCrae, R. R. & John, O. P. 1992. "An Introduction to the Five-Factor Model and Its Applications," *Journal of Personality*, 60(2), 175.

McCrink, K. & Spelke, E. S. 2016. "Non-Symbolic Division in Childhood," *Journal of Experimental Child Psychology*, 142, 66.

McCrory, D. 2010. *Nature's Interpreter: The Life and Times of Alexander von Humboldt* (Cambridge, UK: Lutterworth Press).

McEvoy, P. & Plant, R. 2014. "Dementia Care: Using Empathic Curiosity to Establish the Common Ground That Is Necessary for Meaningful Communication," *Journal of Psychiatric and Mental Health Nursing*, 21, 477.

McGilvray, J. (ed.). 2005. *The Cambridge Companion to Chomsky* (Cambridge, UK: Cambridge University Press).

McMurrich, J. P. 1930. *Leonardo da Vinci, the Anatomist (1452–1519)* (Baltimore: Williams & Wilkins).

Mikulincer, M. 1997. "Adult Attachment Style and Information Processing: Individual Differences in Curiosity and Cognitive Closure," *Journal of Personality and Social Psychology*, 72(5), 1217.

Mirolli, M. & Baldassarre, G. 2013. "Functions and Mechanisms of Intrinsic Motivations: The Knowledge versus Competence Distinction," in *Intrinsically Motivated Learning in Natural and Artificial Systems*, ed. G. Baldassarre & M. Morelli (Heidelberg: Springer).

Mlodinow, L. 2015. *The Upright Thinkers: The Human Journey from Living in Trees to Understanding the Cosmos* (New York: Pantheon).

Moro, A. 2008. *The Boundaries of Babel: The Brain and the Enigma of Impossible Languages*, trans. I. Caponigro & D. B. Kane (Cambridge, MA: MIT Press).

Muentener, P., Bonawitz, E., Horowitz, A., & Schulz, L. 2012. "Mind the Gap: Investigating Toddlers' Sensitivity to Contact Relations in Predictive Events," *PLOS ONE*, 7(4), e34061.

Muniz, V. 2005. *Reflex: A Vik Muniz Primer* (New York: Aperture).

Murayama, K. & Kuhbandner, C. 2011. "Money Enhances Memory Consolidation—But Only for Boring Material," *Cognition*, 119, 120.

Nabokov, V. 1990. *Bend Sinister* (New York: Vintage International).

Neisser, V. (ed.). 1998. *The Rising Curve: Long-Term Gains in IQ and Related Measures* (Washington, DC: American Psychological Association).

Nuland, S. B. 2000. *Leonardo da Vinci: A Life* (New York: Viking).

Nunberg, H. 1961. *Curiosity* (New York: International Universities Press).

O'Connor, D. K. 2014. "Aristotle: Aesthetics," in *Routledge Companion to Ancient Philosophy.* Eds. J. Warren & F. Sheffield (New York: Routledge). p. 387.

Ollman, A. 2016. *Vik Muniz* (Munich: DelMonico Books).

Open Science Collaboration. 2015. "Estimating the Reproducibility of Psychological Science," *Science*, 349, aac4716.

O'Shea, M. 2005. *The Brain: A Very Short Introduction* (Oxford: Oxford University Press).

Otero, C. (ed.). 1994. *Noam Chomsky: Critical Assessments*, vols. 1–4 (London: Routledge).

Oudeyer, P.-Y. & Kaplan, F. 2007. "What Is Intrinsic Motivation? A Typology of Computational Approaches," *Front. Neurobot.*, 1, 6.

Paloyelis, Y., Asherson, P., Mehta, M. A., Faraone, S. V., & Kuntsi, J. 2010. "DATI and COMT Effects on Delay Discounting and Trait Impulsivity in Male Adolescents with Attention Deficit/Hyperactivity Disorder and Healthy Controls," *Neuropsychopharmacology*, 1.

Paloyelis, Y., Mehta, M. A., Faraone, S. V., Asherson, P., & Kuntsi, J. 2012. "Striatal Sensitivity during Reward Processing in Attention Deficit/Hyperactivity Disorder," *Journal of the American Academy of Child & Adolescent Psychiatry*, 51(7), 722.

Pedretti, C. 1957. *Leonardo da Vinci: Fragments at Windsor Castle from the Codex Atlanticus* (London: Phaidon).

Pedretti, C. 1964. *Leonardo da Vinci on Painting: A Lost Book* (*Libro A*) (Berkeley: University of California Press).

Pedretti, C. 2005. *Leonardo da Vinci* (Charlotte, NC: Taj Books International)

Peters, O. (ed.). 2014. *Degenerate Art: The Attack on Modern Art in Nazi Germany 1937* (Munich: Prestel).

Petrosky, T. 2003. "Obituaries: Ilya Pregogine," *SIAM News*, 36(7), https://www.siam.org/pdf/news/352.pdf.

Pevsner, J. 2014. "Leonardo da Vinci, Neuroscientist," *Scientific American: Mind*, 23(1), 48.

Pinker, S. 1994. *The Language Instinct: How the Mind Creates the Gift of Language* (New York: William Morrow).

Pinker, S. 1997. *How the Mind Works* (New York: Norton).

Piotrowski, J. T., Litman, J. A., & Valkinburg, P. 2014. "Measuring Epistemic Curiosity in Young Children," *Infant and Child Development*, 23, 542.

Plomin, R. 1999. "Genetics and General Cognitive Ability," *Nature*, 402 (6761 suppl.), C25.

Plomin, R., et al. 2012. *Behavioral Genetics*, 6th edition (London: Worth).

Povinelli, D. J. & Dunphy-Lelii, S. 2001. "Do Chimpanzees Seek Explanations? Preliminary Comparative Investigations," *Can. J. Exp. Psychol.*, 55(2), 185.

Power, C. 2000, "Secret Language Use at Female Initiation: Bounding Gossiping Communities," in *The Evolutionary Emergence of Language: Social Function and the Origins of Linguistic Form*, ed. C. Knight, M. Studdert-Kennedy, & J. R. Hurford (Cambridge, UK: Cambridge University Press).

Randall, L. 2013. *Higgs Discovery: The Power of Empty Space* (New York: Harper Collins).

Randall, L. 2015. *Dark Matter and the Dinosaurs: The Astounding Interconnectedness of the Universe* (New York: Ecco).

Rappaport, R. 1999. *Ritual and Religion in the Making of Humanity* (Cambridge, UK: Cambridge University Press).

Redgrave, P., et al. 2008. "What Is Reinforced by Phasic Dopamine Signals?," *Brain Res. Rev.*, 58, 322.

Rees, M. 2003. *Our Final Hour* (New York: Basic Books).

Reti, L. 1972. *The Library of Leonardo Da Vinci* (Pasadena, CA: Castle Press).

Richard, J. M. & Berridge, K. C. 2011. "Nucleus Accumbens Dopamine/Glutamate Interaction Switches Modes to Generate Desire versus Dread: D_1 Alone for Appetitive Eating but D_1 and D_2 Together for Fear," *Journal of Neuroscience*, 31(36) 12866.

Richter, I. A. (ed.). 1952. *The Notebooks of Leonardo da Vinci* (New York: Oxford University Press).

Richter, J. P. 1883. *The Literary Works of Leonardo da Vinci* (London: Simpson Low, Marston Searle & Rivington).

Richter, J. P. (ed.). 1970. *The Notebooks of Leonardo Da Vinci* (Mineola, NY: Dover).

Riesen, J. M. & Schnider, A. 2001. "Time Estimation in Parkinson's Disease: Normal Long Duration Estimation Despite Impaired Short Duration Discrimination," *J. Neurol*, 248(1), 27.

Rigol, R. M. 1994. "Fairy Tales and Curiosity: Exploratory Behavior in Literature for Children or the Futile Attempt to Keep Girls from the Spindle," in *Curiosity and Exploration*, ed. H. Keller, K. Schneider, & B. Henderson (Berlin: Springer Verlag).

Risko, E. F., Anderson, N. C., Lanthier, S., & Kingstone, A. 2012. "Curious Eyes: Individual Differences in Personality Predict Eye Movement Behavior in Scene-Viewing," *Cognition*, 122, 86.

Rossing, B. E. & Long, H. B. 1981. "Contributions of Curiosity and Relevance to Adult Learning Motivation," *Adult Education*, 32(1), 25.

Roth, G. & Dicke, U. 2005. "Evolution of the Brain and Intelligence," *Trends in Cognitive Sciences*, 9(5), 250.

Ruggeri, A. & Lombrozo, T. 2015. "Children Adapt Their Questions to Achieve Efficient Search," *Cognition*, 143, 203.

Ryan, R. & Deci, E. 2000. "Intrinsic and Extrinsic Motivation: Classical Definitions and New Directions," *Contemp. Educ. Psychol.*, 25, 54.

Saab, B. J., et al. 2009. "NCS-1 in the Dentate Gyrus Promotes Exploration, Synaptic Plasticity, and Rapid Acquisition of Spatial Memory," *Neuron*, 63(5), 643.

Schacter, D. L., Gilbert, D. T., Wegner, D. M., & Nock, M. K. 2014. *Psychology*, 3rd edition (New York: Worth).

Schewe, P. F. 2013. *Maverick Genius: The Pioneering Odyssey of Freeman Dyson* (New York: Thomas Dunne Books).

Schilpp, P. (ed.). 1949. *Albert Einstein: Philosopher-Scientist* (Evanston, IL: Library of Living Philosophers).

Schulz, L. 2012. "The Origins of Inquiry: Inductive Inferences and Exploration in Early Childhood," *Trends in Cognitive Sciences*, 16, 382.

Schulz, L. E. & Bonawitz, E. B. 2007. "Serious Fun: Preschoolers Engage in More Exploratory Play When Evidence Is Confounded," *Developmental Psychology*, 43(4), 1045.

Shohamy, D. & Adcock, R. A. 2010. "Dopamine and Adaptive Memory," *Trends in Cognitive Sciences*, 14, 464.

Shutts, K., et al. 2011. "Race Preferences in Children: Insights from South Africa," *Developmental Science*, 14:6, 1283.

Siegal, N. 2014. *The Anatomy Lesson*, (New York: Nan A. Talese).

Silvia, P. J. 2006. *Exploring the Psychology of Interest* (Oxford: Oxford University Press).

Silvia, P. J. 2012. "Curiosity and Motivation," in *The Oxford Handbook of Human Motivation*, ed. Richard M. Ryan (Oxford: Oxford University Press).

Singh, S. 1997. *Fermat's Enigma: The Epic Quest to Solve the World's Greatest Mathematical Problem* (New York: Walker).

Sluckin, W., Colman, A. M., & Hargreaves, D. J. 1980. "Liking Words as a Function of the Experienced Frequency of Their Occurrence," *British Journal of Psychology*, 71, 163.

Spielberger, C. D. & Starr, L. M. 1994. "Curiosity and Exploratory Behavior," in *Motivation: Theory and Research*, ed. H. F. O'Neal Jr. & M. Drillings (Hillsdale, NJ: Erlbaum).

Stalnaker, T. A., Cooch, N. K., & Schoenbaum, G. 2015. "What the Orbitofrontal Cortex Does Not Do," *Nature Neuroscience*, 18, 620.

Stephens, J. 1912. *The Crock of Gold* (London: Macmillan), babel.hathitrust.org/cgi/pt?id=mdp.39015031308953;view=1up;seq21.

Steudel-Numbers, K. L. 2006. "Energetics in Homo Erectus and Other Early Hominins: The Consequences of Increased Lower-Limb Length," *Journal of Human Evolution*, 51, 445.

Stringer, C. 2011. *The Origin of Our Species* (London: Allen Lane).

Sykes, C. (ed.). 1994. *No Ordinary Genius: The Illustrated Richard Feynman* (New York: Norton).

Tallerman, M. & Gibson, K. R. (eds.). 2012. *The Oxford Handbook of Language Evolution* (Oxford: Oxford University Press).

Tan, S. J., et al. 2014. "Plasmonic Color Palettes for Photorealistic Printing with Aluminum Nanostructures," *Nano Lett.*, 14(7), 4023.

Tavor, I., et al. 2016. "Task-Free MRI Predicts Individual Differences in Brain Activity During Task Performance," *Science*, 352(6282), 216.

Tomkins, S. 1998. *The Origins of Humankind, Social Biology Topics* (Cambridge, UK: Cambridge University Press).

Unger, R. 2004. *False Necessity: Anti-Necessitarian Social Theory in the Service of Radical Democracy*, revised edition (London: Verso).

Van Arsdale, A. P. 2013. "Homo Erectus—A Bigger, Smarter, Faster Hominin Lineage," *Nature Education Knowledge*, 4(1), 2.

Van den Heuvel, M. P., et al. 2009. "Efficiency of Functional Brain Networks and Intellectual Performance," *Journal of Neuroscience*, 29(23), 7619.

van Veen, V., Cohen, J. D., Botvinick, M. M., Stenger, V. A., & Carter, C. S. 2001. "Anterior Cingulate Cortex, Conflict Monitoring, and Levels of Processing," *Neuroimage*, 14, 1302.

Vasari, G. 1986. *The Great Masters*, trans. Gaston Du C. de Vere (Fairfield, CT: Hugh Lauter Levin Associates).

von Humboldt, A. 1997. *Cosmos: A Sketch of the Physical Description of the Universe*, trans. E. C. Otté, introduction by N. A. Rupke, vols. 1 & 2 (Baltimore: Johns Hopkins University Press). Originally published in 1849.

Voss, J. L., Gonsalves, B. D., Federmeier, K. D., Tranel, D., & Cohen, N. J. 2011. "Hippocampal Brain-Network Coordination During Volitional Exploratory Behavior Enhances Learning," *Nature Neuroscience*, 14(1), 115.

Wang, L., Uhrig, L., Jarroya, B., & Dehaene, S. 2015. "Representation of Numerical and Sequential Patterns in Macaque and Human Brains," *Curr. Biol.*, 25(15), 1966.

Watts Smith, T. 2015. *The Book of Human Emotions: An Encyclopedia of Feeling from Anger to Wanderlust* (London: Profile Books).

White, M. 2000. *Leonardo: The First Scientist* (London: Little, Brown).

White, R. W. 1959. "Motivation Reconsidered: The Concept of Competence," *Psychology Review*, 66(5), 297.

Wilczek, F. 2015. *A Beautiful Question: Finding Nature's Deep Design* (New York: Penguin Press).

Wills III, H. 1985. *Leonardo's Dessert: No Pi* (Reston, VA: National Council of Teachers of Mathematics).

Wilson, J. D. 1987. *A Reader's Guide to the Short Stories of Mark Twain* (Boston: G. K. Hall).

Wilson, T. D., Centerlar, D. B., Kermer, D. A., & Gilbert, D. T. 2005. "The Pleasure of Uncertainty: Prolonging Positive Moods in Ways People Do Not Anticipate," *Journal of Personality and Social Psychology*, 88(1), 5.

Winkler-Rhoades, N., Carey, S., & Spelke, E. S. 2013. "Two-Year-Old Children Interpret Abstract, Purely Geometric Maps," *Developmental Science*, 16, 365.

Wittman, B. C., Dolan, R. J., & Düzel, E. 2011. "Behavioral Specifications of Reward-Associated Long-Term Memory Enhancement in Humans," *Learning and Memory*, 18, 296.

Wolfe, T. 1998. *A Man in Full* (New York: Farrar, Straus & Giroux).

Wood, A. C., Rijsdijk, F., Asherson, P., & Kuntsi, J. 2011. "Inferring Causation from Cross-Sectional Data: Examination of the Causal Relationship Between Hyperactivity-Impulsivity and Novelty Seeking," *Frontiers in Genetics*, 2, article 6, 1.

Wootton, D. 2015. *The Invention of Science: A New History of the Scientific Revolution* (New York: HarperCollins).

Wrangham, R. W. 2009. *Catching Fire: How Cooking Made Us Human* (New York: Basic Books).

Wundt, W. M. 1874. *Grundzüge der Physiologischen Psychologie* (Leipzig: Engelmann).

Yousafzai, M. & Lamb, C. 2013. *I Am Malala: The Girl Who Stood Up for Education and Was Shot by the Taliban* (Boston: Little, Brown).

Zeldin, T. 1994. *An Intimate History of Humanity* (London: Sinclair-Stevenson).

Zhou, C., Wang, K., Fan, D., Wu, C., Lin, D., Lin, Y., & Wang, E. 2015. "An Enzyme-Free and DNA-Based Feynman Gate for Logically Reversible Operation," *Chem. Commun.* 28, 51(51); 10284.

Zöllner, F. 2007. *Leonardo da Vinci: The Complete Paintings and Drawings* (Köln: Taschen).

Zubov, V. P. 1968. *Leonardo da Vinci*, trans. D. H. Kraus (Cambridge, MA: Harvard University Press).

Zuckerman, M. 1984. "Sensation Seeking: A Comparative Approach to a Human Trait," *Behavioral Brain Science*, 7, 413.

Zuckerman, M., Eysenck, S. B. G., & Eysenck, H. J. 1978. "Sensation Seeking in England and America: Cross-Cultural, Age, and Sex Comparisons," *Journal of Consulting and Clinical Psychology*, 46, 139.

Zuckerman, M. & Litle, P. 1985. "Personality and Curiosity about Morbid and Sexual Events," *Personality and Individual Differences*, 7(1), 49.

Zuss, M. 2012. *The Practice of Theoretical Curiosity* (Dordrecht: Springer).

Credits

The author and publisher gratefully acknowledge permission to reprint the following material:

Figure 1: H. A. Weaver, T. E. Smith, STScI, NASA/ESA.

Figure 2: By J. Bedke, STScI, NASA.

Figure 3: Housed in the Mauritshuis Museum in The Hague, Image in the Public Domain.

Figure 4: By Hubble Space Telescope Comet Team and NASA.

Figure 5: Mural painting in the refectory of the Convent of Santa Maria delle Grazie in Milan. Image in the Public Domain.

Figure 6: RCIN 912283. By permission of Royal Collection Trust/©Her Majesty Queen Elizabeth II 2016.

Figure 7: In the National Gallery of Art, Washington, D.C., Alisa Yellon Bruce Fund. Image in the Public Domain.

Figure 8: By permission of Royal Collection Trust/©Her Majesty Queen Elizabeth II 2016.

Figure 9: In the collection of Tate Britain. Image in the Public Domain.

Figure 10: By Joseph Weber. Reproduced with permission of Virginia Trimble.

Figure 11: Page with drawings from Feynman's sketchbook from 1985. In Feynman 1995b. Courtesy of Museum Syndicate.

Figure 12: From Leonardo da Vinci's *Codex Atlanticus*, at the Biblioteca Ambrosiana, Milan. By permission of Getty Images.

Figures 13–16: By Paul Dippolito

Figure 17: Courtesy of Elizabeth Bonawitz.

Figure 18: By Paul Dippolito

Figure 19: From Jepma et al. 2012. Reproduced with permission of Marieke Jepma.

Figure 20: From Herculano-Houzel 2009. Reproduced with permission of Suzana Herculano-Houzel.

Figure 21: "Lucy" skeleton (AL 288-1), *Australopithecus afarensis*, cast from Museum National d'Histoire Naturelle, Paris. Image in the Public Domain.

Figure 22: Photo courtesy of Katie Reisz.

Figure 23: An image of Richard Feynman by Vik Muniz, from the "ink series." By permission of Vik Muniz.

Figure 24: The smaller of the two Buddhas of Bamiyan in 1977. Image in the Public Domain.

Figure 25: Photo taken by Yee Ming Tan in 2008.

Good faith efforts have been made to contact the copyright holders of the art and text in this book, but in a few cases the author has been unable to locate them. Such copyright holders should contact Simon & Schuster, 1230 Avenue of the Americas, New York, NY 10020.

Index

Page numbers in *italics* refer to illustrations.

ALSO BY MARIO LIVIO

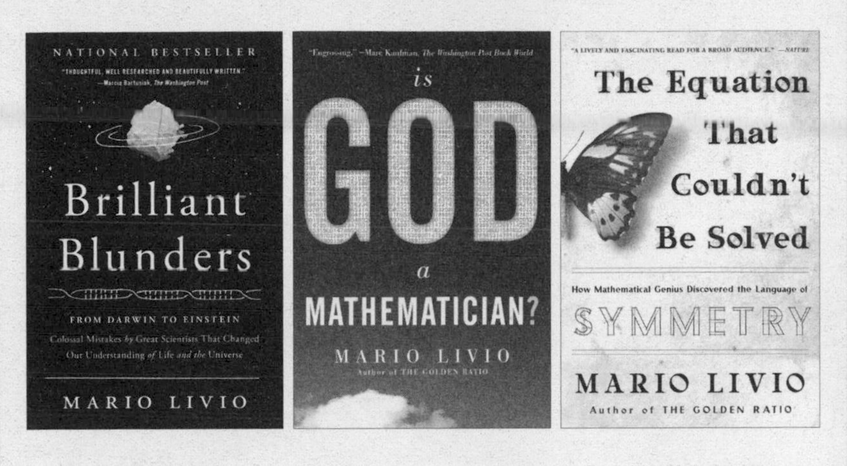

"Thanks to his deep curiosity, Livio turns *Brilliant Blunders* into a thoughtful meditation on the course of science itself."
—CARL ZIMMER, *THE NEW YORK TIMES BOOK REVIEW*

"With elegance and clarity, Mario Livio charts how, through science and mathematics, we have come to glimpse the fundamental rules on which the universe runs."
—PAUL DAVIES, AUTHOR OF *THE GOLDILOCKS ENIGMA*

"A lively and fascinating read for a broad audience."
—*NATURE*

PICK UP OR DOWNLOAD YOUR COPIES TODAY!

57373

"An energetic look at the psychology and neuroscience of our inquisitiveness."—DAN JONES, *Nature*

• • •

WHY are we more distracted by a cell-phone conversation, where we can hear only one side of the dialogue, than by an overheard argument between two people? Are children more curious than adults? What is the source of the morbid curiosity that causes bystanders to gather at crime scenes or traffic accidents? What evolutionary purpose does curiosity serve? How does our mind choose what to be curious about? *Why?* explores these and many other intriguing questions.

Curiosity is essential to creativity; it is a necessary ingredient in art forms from literature to the visual arts to music. It is the principal driver of science, and yet there is no scientific consensus on why we humans are so curious or about the precise mechanisms in our brain that are responsible for curiosity. In *Why?*, insatiably curious and bestselling author Mario Livio investigates this very human phenomenon in an irresistible and entertaining book that will captivate anyone who is curious about curiosity.

"Filled with fascinating stories, tidbits, and psychological insights, *Why?* is a delightful romp through every aspect of human curiosity. It will surprise you, make you smarter, and put a spring in your step." —**STEVEN STROGATZ, Jacob Gould Schurman Professor of Applied Mathematics, Cornell University, and author of *The Joy of X***

• • •

MARIO LIVIO is an internationally known astrophysicist, a bestselling author, and a popular speaker. He has written five previous books and more than 400 scientific articles, has appeared on television programs ranging from *60 Minutes* and *NOVA* to *The Daily Show with Jon Stewart*, and has lectured at venues such as the Smithsonian Institution, the Royal Astronomical Society, and TEDxMidAtlantic.

Visit the author at WWW.MARIO-LIVIO.COM

PSYCHOLOGY 0518

ISBN 978-1-4767-9210-1 **$16.00 U.S.**/$22.00 Can.

9 781476 792101 51600

PRINTED IN THE U.S.A.

COVER ILLUSTRATION BY JASON HEUER